LOVE AND INTIMACY IN CONTEMPORARY SOCIETY

Love and Intimacy in Contemporary Society reflects on relationships in contemporary society and the role of love and intimacy in framing lives. The book draws on sociological perspectives, cultural sociology and gender theory perspectives.

It looks at how love and intimacy is experienced differently and intersected by gender, ethnicity, race and sexuality. This book aims to encourage people to understand theories of intimacy, emotions and desire by examining these concepts contemporaneously and cross-culturally. It also explores how love and intimacy is experienced by young people and how it is impacted by age. It looks at its representation in the media and film and focuses on how gender, ethnicity and sexuality offer different perspectives on love and intimacy.

The book shows how relationships are impacted by social networking and new technologies and the opportunities and challenges posed by these new platforms for building relationships. Finally, the book examines how intimacy has become commercialised in late capitalism and how that acts to change relationships. The book is written in an accessible way and explores a range of theoretical debates and contemporary research around emotions, which can be useful for undergraduate, postgraduate and doctoral study.

Ann Brooks is a Visiting Professor at the Australian Catholic University in 2018–2020. She is a Fellow of the Academy of Social Sciences (FAcSS). Her latest books are *Genealogies of Emotions, Intimacy and Desire: Theories of Changes in Emotional Regimes from Medieval Society to Late Modernity* (2017) and *Women, Politics and the Public Sphere* (2019).

"Senior international scholar Ann Brooks delivers a valuable contribution to the fast-growing areas of critical love studies and sociology of emotions with this interdisciplinary look at love. Her cutting-edge survey of current research provides a 'state-of-the-field' framework examining relationships, globalization, today's online culture, and more. A masterly synthesis!"

> — **Catherine M. Roach**, Professor of Gender and Culture Studies, University of Alabama; Author of *Happily Ever After: The Romance Story in Popular Culture*, and the novels *Master of Love* and *Knight of Love*

"The scope and diversity of scholarship on love and intimacy can be daunting, but Ann Brooks's *Love and Intimacy in Contemporary Society: Love in an International Context* offers an accessible, richly informative introduction. Anyone interested in real-life love and/or its media representations will find much to learn."

> — **Eric Murphy Selinger**, Professor of English at DePaul University; Editor of the *Journal of Popular Romance Studies*, author of *What Is It Then Between Us? Traditions of Love in American Poetry*, and co-editor of *Romance Fiction and American Culture: Love as the Practice of Freedom*

LOVE AND INTIMACY IN CONTEMPORARY SOCIETY

Love in an International Context

Ann Brooks

Routledge
Taylor & Francis Group

LONDON AND NEW YORK

First published 2020
by Routledge
2 Park Square, Milton Park, Abingdon, Oxon OX14 4RN

and by Routledge
52 Vanderbilt Avenue, New York, NY 10017

Routledge is an imprint of the Taylor & Francis Group, an informa business

Routledge and the Author acknowledge permission given by the
Museum of Contemporary Art, Sydney, for the reproduction of the
original artwork by Sanne Mestrom, *Soft Kiss*, 2011, Museum of
Contemporary Art, Sydney, Australia

British Library Cataloguing-in-Publication Data
A catalogue record for this book is available from the British Library

Library of Congress Cataloging-in-Publication Data
Names: Brooks, Ann, 1952– author.
Title: Love and intimacy in contemporary society : love in an
    international context / Ann Brooks.
Description: Abingdon, Oxon ; New York, NY : Routledge, 2020. |
    Includes bibliographical references and index.
Identifiers: LCCN 2019029763 (print) | LCCN 2019029764
    (ebook) | ISBN 9781138572935 (hardback) | ISBN 9781138572331
    (paperback) | ISBN 9780203702123 (ebook)
Subjects: LCSH: Love—Cross-cultural studies. | Intimacy
    (Psychology)—Cross-cultural studies. | Interpersonal
    relations—Cross-cultural studies. | Globalization—Social
    aspects. | Technological innovations—Social aspects.
Classification: LCC GT2630 .B76 2020 (print) | LCC GT2630
    (ebook) | DDC 306.7—dc23
LC record available at https://lccn.loc.gov/2019029763
LC ebook record available at https://lccn.loc.gov/2019029764

ISBN: 978-1-138-57293-5 (hbk)
ISBN: 978-1-138-57233-1 (pbk)
ISBN: 978-0-203-70212-3 (ebk)

Typeset in Bembo
by Apex CoVantage, LLC

CONTENTS

AUTHOR BIOGRAPHY

Ann Brooks is a Visiting Professor at the Australian Catholic University (Melbourne and Sydney), at the Institute of Religion, Politics and Society in 2018–20. She is a fellow of the Academy of Social Sciences (FAcSS). She has held senior academic positions in universities in Australia, Singapore, the UK and New Zealand. Most recently, she was Professor of Sociology at Bournemouth University. She has also held research fellowships at the University of California, Berkeley, and the Asia Research Institute at the National University of Singapore. Ann has been an international research investigator with the Australia Research Council funded Centre of Excellence for the History of Emotions 2011–2018 in Australia. She is author of *Academic Women* (Open University Press, 1997); *Postfeminisms: Feminism, Cultural Theory and Cultural Forms* (Routledge, 1997); *Gender and the Restructured University* (Open University Press, 2001); *Gendered Work in Asian Cities: The New Economy and Changing Labour Markets* (Ashgate, 2006); *Social Theory in Contemporary Asia* (Routledge, 2010); *Gender, Emotions and Labour Markets: Asian and Western Perspectives* (Routledge, 2013) and *Emotions in Transmigration: Transformation, Movement and Identity* (Palgrave, 2012) (with Ruth Simpson); *Popular Culture, Global Intercultural Perspectives* (Palgrave, 2014); *Consumption, Rights and States: Comparing Global Cities in Asia and the US* (Anthem Press, 2014) (with Lionel Wee); and *Emotions and Social Change: Historical and Sociological Perspectives* (Routledge, New York, 2014) (co-edited with David Lemmings). Her latest books include *Genealogies of Emotions, Intimacy and Desire: Theories of Changes in Emotional Regimes from Medieval Society to Late Modernity* (Routledge New York, 2017) and *Women, Politics and the Public Sphere* (University of Bristol/Policy Press, 2019).

FOREWORD

Romantic love and intimacy in Australia

Australian culture often appears profoundly ambivalent about romance and marriage. On the one hand, we apparently have an insatiable appetite for American romantic comedies, melodramas about love and marriage, love songs and local versions of American romance-based game shows or reality television programs such as *Perfect Match*, *The Bachelor/Bachelorette*, the Danish-based *Married at First Sight*, or the British-derived *The Farmer Wants a Wife* and *First Dates*. These are all local adaptations of foreign concepts that sell or play with romantic love. Original, home-grown examples of romantic love in Australian culture – whether 'high' or popular culture – are rather more difficult to find. While Australian popular culture certainly has examples of love songs, romantic comedies and dramas, and a flourishing romance novel industry, we as a nation are generally more famous for our feminist denunciations of love and marriage. In *The Female Eunuch* (1970), Germaine Greer complained that the concept of romantic love had devastating effects on women, teaching them to cherish 'the chains of their bondage' (Greer 1993: 176). Greer's work triggered a slew of ground-breaking books in Australian feminist scholarship – such as Anne Summers' *Damned Whores and God's Police: The Colonization of Women in Australia* (1975) and Miriam Dixon's *The Real Matilda: Women and Identity in Australia, 1788–1975* (1976) – critically examining the deleterious effects of colonialism, industrialisation and marriage and family life on women in Australia. Only recently has there been a turn towards an exploration of romantic love and intimacy in Australian culture (Maroske 1985; Simmonds 2005; Teo 2005, 2006, 2014, 2017; Grossi 2014, 2015). Mostly, however, the subject of romantic love has been ignored, buried, condemned in scholarship or self-consciously addressed and often mocked in popular culture.

If the reasons for this discomfort are cultural, they are also perhaps historically rooted in the fact that romantic love, sex, intimacy and marriage in Australia have always been entangled with underlying assumptions about political citizenship. As King (2017) and Elder (2017) both observe, who can freely fall in love and who is considered 'marriageable' are questions tied to national belonging and citizenship rights: 'love and marriage often became a plot device for marking who is Australian and who is not' (Elder 2017: 140). It is perhaps this, more than anything else that distinguishes the history of romantic love and intimacy in Australia from their English-speaking counterparts elsewhere. There are dark and disturbing reasons as to why love and marriage provoke a national discomfort when we consider the origins of Australian society.

This essay presents a thesis about the origins and development of ideas about marriage, romantic love and intimacy in Australian history from the colonial era to the present day. It argues that the history of the convict era, with its horrific exploitation and abuse of convict women, and the frontier violence experienced by Aboriginal women, have cast a long shadow on the promise of love, intimacy and marriage that is missing from British history, and under-examined in American scholarship on romantic love. Despite such an unpropitious start, however, the mid-nineteenth century saw a burgeoning culture of romantic love that bears similarities to that of Britain and the United States, as free immigrants from Britain began to settle the colonies. I am particularly intrigued by the way growing American influence in the twentieth century was experienced in the form of romantic consumerism, but how Australian culture also remained, on the whole, resistant to the optimistic imperative of the American popular culture of romantic love (Roach 2016).

The history of white heterosexual romantic love and intimacy in Australia is different from other English-speaking countries, such as Britain or the United States, because of Australia's particular conditions of colonial settlement. As most people know, the east coast of Australia was first settled as a penal colony at Sydney in 1788, followed by the establishment of other penal settlements in Tasmania, Victoria, Queensland and Western Australia. It was not until after the 1820s that free immigrants began to settle in the colonies, contributing to the push by the "respectable" middle classes to end the convict system in Australia. Transportation was progressively ended throughout the different colonies between 1839 and 1868, when the last convict ship arrived in Western Australia.

The convict system created special conditions that made gender relations in the early colony vastly different to social conditions in Britain and the United States at the same time. The percentage of male prisoners far outnumbered female convicts. Of the roughly 164,000 prisoners transported from Britain to Australia between 1788 and 1868, women numbered about 25,000 – less than sixteen per cent of the total convict population (Damousi 1997: 2). This gender imbalance had a significant impact on the history of love, intimacy and marriage in colonial Australia that is unmatched in Europe or America during the same period.

British and American scholars generally argue that the 'love match' – where the potential spouse was freely chosen for love rather than pecuniary or other interests – was becoming more common in Britain (Giddens 1992: 29) and the United States (Rothman 1987: 31; Stone 1988: 19) at the end of the eighteenth and nineteenth centuries. This practice gave rise to the ritual of courtship, as well as changing ideas about love. Where love between husband and wife had was previously defined as care, mutual assistance and cooperation to achieve shared goals, throughout the nineteenth century love became something of a puzzle and a mystery. Countless articles were published in periodicals – 'high-toned intellectual and literary journals as well as in low-quality penny newspapers' (Weisser 2013: 52) – contemplating the meaning of love, falling in love, and the nature of 'true love'. In many ways, this concern with defining romantic love is understandable. The love match increasingly gave young people the power of choice over whom they were to marry, but if this choice were to determine their future happiness and well-being, they needed to know what love really was (Weisser 2013: 55). Interestingly, 'true love' did not necessarily include 'passion' and 'romance' in the early nineteenth century, for these were understood as evanescent and unstable emotions rather than authentic love, and the extreme emotions were therefore mistrusted. Instead, love was considered 'more compelling than friendship, more lasting than passion, more serious than romance' (Rothman 1987: 36).

By the mid-century, however, 'romance' – with all the intense feelings of thrill, excitement, passion, and yearning that this term entailed – had become fused with love in courtship. Romance was now believed to be 'the key to domestic harmony rather than as a threat to it' (Rothman 1987: 103). The nineteenth-century version of romantic love marked a departure from the medieval version in that it was rooted in a notion of a private, autonomous 'ideal self', and the progression of romantic love was marked by a gradual disclosure of this 'ideal self' to the beloved in order to ensure there would be no surprises after marriage (Lystra 1989). Candour, frankness and honest self-disclosure, it was believed, laid the foundations for 'mutuality, commonality, and sympathy between man and woman – precisely those qualities most likely to bridge the widening gap between home and world' (Rothman 1987: 107). In other words, this was the moment when romantic love became fused with expectations of emotional, intellectual and psychological intimacy in Britain and the United States.

In colonial Australia during the convict era, however, there was no such expectation among the majority convict population of romantic love or an emotional or psychological intimacy gradually established in courtship through self-disclosure. As many Australian feminist scholars from Anne Summers (1975) to Joy Damousi (1997) show, gender relations in the early colony at Sydney were characterised by sexual assault and exploitation of convict women by other convict men as well as the naval and military officers in charge of their punishment and servitude, but also their well-being. 'It was no wonder,' Damousi (1997: 35) argues, 'that convict women used sex as a commodity to survive, given the high demand for their services not only as servants and mistresses, but also as prostitutes.' Despite circumstances which left convict women little choice but to seek protection as mistresses or in sex work, these women were regarded as sexual deviants who 'polluted' or 'contaminated' the colony.

In an attempt to clean up the morals of the colony, convicts were encouraged to marry. Convict women certainly thought that marriage would provide 'protection and an escape from destitution' (Damousi 1997: 36). Certainly, there was nothing at all romantic, loving or intimate about the way convict courtship or marriages were initially arranged. Damousi notes that convict courtships were a form of popular entertainment where 'crowds gathered to ogle and catcall' (Damousi 1997: 52) while convict women were brought out for inspection and lined up like cattle in a fair. James Mudie, who arrived in Sydney in the 1820s, described how the male convict then 'goes up and looks at the women, and if he sees a lady that takes his fancy, he makes a motion to her, and she steps on one side' (Damousi 1997: 53). Convict women still had the power of refusal, but few exercised that option given their limited opportunities of survival, let alone advancement, and the dangers of assault within the colony. Of course, even with such unromantic and inauspicious beginnings, there is every possibility that these marriages, like other illicit relationships, could develop into caring, cooperation and love, but Damousi argues that many convict women discovered that 'violence, assault and abuse within marriage was common' (Damousi 1997: 36).

It wasn't only white convict women who suffered in relations at the hands of the settlers – convict and free. White settlement was also accompanied by extreme violence against Indigenous peoples, including the enslavement and sexual abuse of Aboriginal women along frontier and outback regions that continued well into the early twentieth century. Larissa Behrendt notes that especially in the outback, sexual access to Aboriginal women were one of the promised 'perks' of bush life for settlers, resulting 'gin hunts' and 'gin sprees' where Aboriginal women were forcibly abducted for sex (Behrendt 2000: 353). 'Conflict over women was a constant feature of frontier relations and sexual relations between white men and black women were a major source of misunderstanding, bitterness and conflict' (Behrendt 2000: 354). Despite being outlawed in 1837, the kidnapping of Aboriginal women for sex continued well into the 1930s (McGrath 1984: 256). Aboriginal women sometimes chose to enter 'voluntarily' into longer term sexual relationships with white men for a variety of reasons, including the protection and survival of their families, or 'to escape undesired marriage or tribal punishment or to gain access to the many attractive possessions of the Europeans', but such relationships were nevertheless formed within deeply unequal power relations, and against 'the background of continual frontier and sexual violence' (Behrendt 2000: 354).

Being regarded as 'low-class prostitutes' had long-lasting effects on what Andrew King (2017) calls Aboriginal women's 'marriageability' – a term that refers not only to their sexual attractiveness, but also how white mainstream society regarded Indigenous people as legal marriage partners. For King, the issue of marriageability is important because it not only positions Aborigines as the subjects and agents of the mainstream narrative of romantic love and intimacy – a narrative that has a strong drive towards marriage and the possibility of a 'happy ever after' – but marriageability also normalises the subjecthood or citizenship of Indigenous people. As King shows, the issue of white-Aboriginal marriageability was salient even in the mid-twentieth century, when the marriage of a white man, Mick Daly, to an Aboriginal woman, Gladys Namagu, was challenged in the Northern Territory court despite Mick's open declaration of love for Gladys in court (King 2017: 91). This was despite the fact that for over a century, the political, religious and institutions of the colony, and then the nation, had been pushing for marriage as a 'civilizing' process to 'tame' convict women and to wipe away the stain of the colony's convict heritage (King 2017: 97).

The drive to clean up the morals of the early colony saw a push for more female free immigrants from the early nineteenth century onwards, especially during the 1840s and 1850s when free middle-class women were expected to exert a civilising influence as 'God's police' on the rowdy colonies (Summers 1975), but men still outnumbered women for most of the century. Marriages were easy to contract, but love and intimacy were harder to achieve. As the authors of *Creating a Nation* (1996: 59–60, 89–92) argue, even for free white women, marriage was not a safeguard against violence and poverty. Feminist scholars such as Penelope Russell are equally dubious about the chances of happy marriages for women in the more prosperous classes, arguing that:

> Against the economic power, masculine strength, sexual freedom and patriarchal dominance of men in society and within the family, women could counterpose only love – a love which they had to attempt to foster and to earn through their physical attractiveness and dutiful, loving behaviour, and which would constrain their husbands to be considerate, non-violent, faithful, and even at times to respect their views and allow them some decision-making power.
>
> *(Russell 1988: 15)*

Australian women's anxieties and fears over marriage were given expression in colonial narratives about love, whether fiction or nonfiction. Whereas the nineteenth century saw the burgeoning of romantic novels in Britain and the United States featuring dramas about courtship and love that, even if they did not always end happily, nevertheless celebrated the ennobling, uplifting effects of true love, the same is not true of colonial Australian love stories. In my research (Teo 2014) into colonial and early Federation love stories, I found that many writers – both male and female – were pessimistic about romantic love in the colonies until the early twentieth century. Love fails to flourish for various reasons, among them: immigrant women's vulnerability en route to Australia; the precarity of their attempts to establish respectable lives in the colony; and the problems created by misogynistic masculinity in a frontier society, where alcoholism exacerbated domestic violence, abandonment of families, or men who renege on the promises made to women during courtship. Alecia Simmonds' (2005) examination of breach-of-promise court cases in colonial Australia from the 1820s until the early twentieth century illuminate two recurrent themes relating to narratives of love and intimacy during this period: firstly, the 'tale of female vulnerability and male duplicity in matters of the heart'; and secondly, the intrusion of the state into personal, private relations as the state attempted to impose the institution of marriage onto a 'recalcitrant' working-class population, using the law to 'disseminate and enforce the norms of Western bourgeois, conjugal love' (100).

With the growth of the colonial middle class from the mid-nineteenth century onwards, the importation of middle-class Victorian concepts of love was inevitable, and it is at this point that the history of romantic love in Australia begins to converge with its British and American counterparts. Like their counterparts in other English-speaking countries, middle-class Australians in the mid-nineteenth century understood romantic love to be:

> an emotional, moral, physical and spiritual attraction believed to be a necessary prerequisite to courtship, with companionate marriage as its ideal goal. . . . Love was supposed to have an ennobling, morally and spiritually uplifting effect, especially upon the male lover.
>
> *(Teo 2006: 177).*

Yet one crucial difference during this period was that where Karen Lystra argues that nineteenth-century Americans tended to regard love as something mysterious or mystical, Australians were on the whole more prosaic and pragmatic about love and intimacy. Sarah Maroske's (1985) and Penelope Russell's (2017) research shows that concepts of love among Evangelical middle-class colonials was initially based on the same discourse of esteem for character and distrust for the 'pretty' and 'flattering' conventions of romantic feelings. Russell goes further in arguing that feelings of attraction and love were tempered by 'pragmatism, even self-interest' during the mid-nineteenth century. The courting lovers Russell examines schooled their feelings so that the marriages they contracted were 'within the acceptable limits of church, business and family' (2017: 44). It is not that these middle-class men and women and women did not feel romantic love; the affect displayed in their letters is a testament to attraction, hope, desire, disappointment and pain. Love was a 'messy, lived experience', Russell observes; it could 'be both opportunistic and sincere – both spiritual and sensual – both familial and individual'. For both the religious and non-religious middle-class in colonial Australia, where the social mobility of emancipated convicts and free working-class immigrants who quickly accrued wealth created unstable social hierarchies, maintaining social status through marriage was important. Romantic love and courtship were therefore refracted through 'correct manners and behaviour, including the ways in which romantic feelings were communicated' (Bailey 2017: 70). Middle-class Australian lovers were aspirational in every way: they sought personal, social and economic progress, as well as moral perfectibility for themselves and their lovers.

What we see in love letters written between middle-class courting couples from the mid-nineteenth to the early twentieth century is a gradual relaxation of formality and manners, and a transition from the language of esteem to the language of passion and intimacy, in consonance with developments in other English-speaking countries. When John proposed to his future wife Ellen in 1867, his passion was subordinated to his esteem for her character: '[Y]our excellent qualities have long been under my silent observation', he wrote to her.

> The time has arrived when, possessed of sufficient to make home happy, and with bright prospects before me, I feel that I may safely and even properly seek for an [sic] helpmate.
> How ardently do I desire to be partaker of that blessing which I am persuaded must ensue from an union with one who is a loving daughter, an affectionate sister and an earnest and sincere friend!
>
> *(Teo 2005: 348)*

The couple had not achieved intimacy in the sense of sharing their deepest, most private 'selves' with each other in an act of mutual self-disclosure, as was common with the white middle-class letter writers examined by Karen Lystra (1989). Rather, intimacy was a condition of close proximity, resting on how well Ellen already fulfilled her family and social roles. By the end of the century, however, formality began to give with to a more casual familiarity between courting couples, and with this came the establishment of intimacy defined as the disclosure of the self, as was common in American examples of courtship during this period (Rothman 1987; Lystra 1989).

American scholars (Bailey 1988; Rothman 1987; Seidman 1991) suggest that the practice of self-disclosure to establish intimacy within romantic love declined during the early twentieth century, as 'old-fashioned' courtship gave way to 'modern' dating, and as the understanding of romantic love became increasingly sexualised and became associated with pleasure, rather than with older medieval notions of quest, suffering and sacrifice. That romantic love should involve less self-disclosure creating mutual understanding in the early twentieth century is ironic, since courtship took place under surveillance: lovers were constantly under the gaze of family and friends since courtship usually took place within the domestic sphere, or in public venues such as churches or community-organised dances. In the early twentieth century, courtship increasingly involved dating: going out together in public places such as tea rooms, dance halls, cabarets, restaurants, cinemas, and other similar activities. Although these spaces were public, especially in urban centres, dating couples could enjoy anonymity and 'islands of privacy' (Illouz 1997: 56), which left them free to experiment sexually. 'Intimacy' reverted to its older meaning of sexual intimacy, since the venues of dating that 'lent themselves to sexual experimentation but not to emotional openness. Dancing cheek to cheek or sitting side by side in a darkened nickelodeon may have invited new physical freedom, but neither encouraged the heart-to-heart talking that had occupied couples in earlier generations' (Rothman 1987: 225).

My study (Teo 2005) of Australian love letters reveals that while Australians also began to engage in similar practices of dating in the early twentieth century, and while managing the effects of their visual appearance was particularly important to the young flappers Liz Conor (2004) studies, there was no corresponding diminution in intimacy or expectations of self-disclosure among lovers as had happened in the United States. In fact, Australian letter writers from both the middle and working classes became increasingly frank about their lives and bodily experiences throughout the twentieth century, especially as understandings of love became increasingly sexualised during this period in accordance with this development in the United States. Men and women discussed their virginity, their sexual experience or lack thereof, and exchanged ideas about how to rectify this before marriage (Teo 2005: 356–357). Australian women controlled the degree of details about sex and other types of behaviour disclosed between them and their lovers, and they hence controlled the level of intimacy afforded to men during courtship. During the 1940s, working-class women writers frankly confessed to binge drinking, wild nights out with girlfriends, stormy family arguments where they themselves had behaved badly, and men in turn confessed to various flaws, bodily functions, and even unheroic behaviour during the war (Teo 2005: 354–355). Where Americans involved in romantic dating in the first half of the twentieth century felt the pressure of trying to attract someone and to maintain their interest, to be a fun, enjoyable and interesting 'date', and who therefore put an increasing emphasis on physical appearance, clothing, 'personality', and the thrill of sexual passion rather than nineteenth-century frankness, candour and self-disclosure (Rothman 1987: 225–226), working-class Australian lovers appeared to share their lives, warts and all, and to insist that their lovers give them unconditional love and acceptance.

> There was no sense, on the part of these lovers, that they had a certain "image" or "style" or "line" (Rothman 1987: 225) to maintain; they fully expected their loved ones to understand their flaws and continue to support and love them regardless. Indeed, the demand for understanding and support in light of these confessions became the very proof of romantic love in the twentieth century.
>
> *(Teo 2005: 354–355).*

Yet as practices of dating became more common, the understanding of romantic love in Australia began to change in one very distinctly gendered way as romance became intertwined with consumerism. Eva Illouz (1997) argues that as Americans began to replace courtship with dating in the early twentieth century, the end-goal of dating ceased to be marriage and began to focus more on pleasure – particularly the pleasure associated with sexual activity and consumption during dates. Dating taught Americans to commodify each other, and it conferred upon romance an exchange value that was reinforced by advertising. Eventually, 'romance' would come to be associated with gifts or experiences that could be bought: boxes of chocolates, perfume, candlelit dinners in expensive restaurants, cosying up together in sophisticated cocktail lounges, cruises at sunset, and romantic holidays. Advertisements for 'ego-expressive' products such as shampoo, powder, toothpaste, deodorant, aftershave and so forth reinforced sex attractiveness with romance and consumerism because the romantic couple, or suggestions of a 'happy ever after' message were used to sell such products. This narrative of romantic consumption was reinforced in mainstream American culture through film, pop songs, romance novels and magazines, which all commodified romance and romanticised commodities.

This American culture of romantic consumerism was imported into Australian women's magazines from the 1920s onwards. 'The impetus towards Americanisation in Australian advertising styles and images . . . was driven by the perception of American women's modernity and the glamour of romantic consumption' (Teo 2006: 181). In adopting Americanised practices of advertising that associated consumption with romance, Australian women's magazines were decades ahead of men's magazines. It wasn't until the 1960s, when American magazines such as *Playboy* began to be imported, that Australian men were 'introduced to a culture of romanticised (and, of course, sexualised) consumption for all sorts of products' (Teo 2006: 185).

This disjunction in how the concept of romance and consumerism was understood by men and women did not have any major implications until American servicemen were stationed in Brisbane and Sydney during the Second World War, and Australian men began to perceive these GIs as sexual and romantic competitors. Marilyn Lake (1990, 1992) and Lyn Finch (1995) have explored the backlash against Australian women during the war as young women were condemned for preferring to date American soldiers over Australian men. Finch (1995: 110) argues that 'American troops, raised in the very centre of consumption capitalism, brought to Australia cultural baggage that included a clear concept of the correlation between spending power and sexual attractiveness'. This nexus between sexual attraction, dating, consumption and 'falling in love' was something that Australian women understood and responded to during the war. I am not suggesting that Australian women did not continue to fall in love and have romantic relationships with Australian men during the war. Rather, I am arguing that for some women, American soldiers – in addition to the glamour of a familiar but foreign culture refracted through Hollywood, and also their better pay – could perform romance in a way that was more attractive and compelling (Teo 2006: 186–191), with the result that many Australian women married American men and moved to the United States at the end of the war (Arrowsmith 2013).

By the post-war years, the 'companionate' marriage based on romantic love tended to be the norm in Australia, as it was also in Britain and the United States. Yet, as Andrew Cherlin (2010) observes, mid-twentieth century expectations of the love match were still lower than what would later follow. Liking, sexual satisfaction and the fulfilment of complementary, gender-defined duties in marriage seemed enough. The importance of sexual intimacy in sustaining romantic love can be gauged by the number of sexual studies and marriage manuals that emphasised the importance of getting sex right. In Australia, Frank Bongiorno (2017: 326) argues, 'Sexual compatibility between a husband and wife became the foundation of a successful marriage. The mutual orgasm was held up as the aim of every married couple, not, as medical opinion once held, because it was necessary for impregnation, but in order to preserve a happy and fulfilling relationship'.

Despite the increased access to and availability of sex, especially after the introduction of the contraceptive pill in 1960 and the sexual revolution of the sixties and seventies, a sexualised understanding of romantic love failed to hold marriages together and divorce rates instead increased. Historians give several reasons for this but generally agree that when marriage is contracted primarily for romantic love, and when romantic love is conceived as something pleasurable that will bring personal fulfilment and even self-transformation, then there is little reason to be condemned to a lifelong marriage if it becomes unfulfilling, especially once sex also became more available outside marriage, and decoupled from romantic love (Langhamer 2013).

Second-wave feminism, with its powerful and cogent critiques of romantic love as a patriarchal tool which kept women in bondage in unequal relationships, was both a symptom of, and a fuel for, this discontent. There could be no possibility of what Marcus Collins (2003: 5) calls 'mutuality' – the 'notion that an intimate equality should be established between men and women through mixing, companionate marriage and shared sexual pleasure' – without equality of the sexes. It is for this reason that Shumway (2003: 139) claims that 'feminism changed intimacy itself', for the movement for equal rights in the economic and domestic sphere enabled what Anthony Giddens calls the 'transformation of intimacy' from relationships built on self-interest, to the 'pure relationship' that 'exists solely for whatever rewards that relationship can deliver' (Giddens 1991: 6).

Increasingly, processes of transnational corporate globalism created parallel cultures of love around the English-speaking world. Novels, self-help books, films, pop songs, and syndicated magazine articles focusing on narratives of love and romance circulated internationally, freighting expectations of love with pleasure and consumption, promising the transformation of the autonomous self as well as the tantalising possibilities of a state of intimacy that would fulfil what was missing in the self. Australia in the late twentieth and twenty-first centuries displays interesting convergences and divergences from an American ideal about love and intimacy that was disseminated globally through the powerful reach of American mass media and popular culture.

Catherine Roach (2016) argues that by the twenty-first century, the United States, at least, was steeped in a contradictory culture of romantic love and intimacy, fuelled by a century of mass-market entertainment wherein the promise of the American Dream could be fulfilled by falling in love. In the United

States, romantic love is a story that gives meaning, coherence and purpose to people's lives:

> According to this story, despite the risks, love is what gives value and depth to life. Our purpose is to bond with a well-suited mate worthy of our love and to love and be loved by this mate within a circle of family and friends. Here, life is a high-stakes quest for the Holy Grail of One True Love. This search is driven by yearning and desire for the paradise of this romantic happily-ever-after. We chase romance, we structure our lives around it, we fashion much of our art and culture from it. The romance story is not only a narrative but becomes also, more disturbingly, an imperative.
>
> *(Roach 2016: 4)*

In a secular, capitalist society that uses sexualised individuation to sell products while reducing individuals to interchangeable commodities, romantic love has become a secular religion in which the 'the redemptive or resurrection power of love' heals all wounds, conquers all, functions 'as a positive force for the good in people's lives' and makes the world a better place (Roach 2016: 23).

While Australians eagerly consume these American narratives of romantic love, intimacy and the promise of a 'happily-ever-after' with your one true love, Australian culture has been far more reticent, and even ambivalent, about articulating such beliefs. Our national literature tends to display a profound pessimism in the redemptive, transformative possibilities of love. The neglect of love as a major theme in Australian literature throughout the twentieth century was such that literary scholars remarked on its absence during a conference in 1983 focusing, paradoxically, on 'The Theme of Love in Australian Writing'. Dorothy Green (1983: 45) argued that Australian literature 'is not rich in the prose or poetry of love in its self-transcendent sense', while the poet Fay Zwicky (1983: 30) remarked: 'Our great novelists . . . though dab hands at loneliness, terror, and human indignity, tend to avoid treating the passionate encounter of a man and a woman which we have come to expect at the centre of the European novel'. Patricia Dobrez (1983: 3) qualified that it is 'not that Australians are incapable of love, that they do not feel it, but that its flow is soon diverted into channels of pessimism and despair'.

While Australia has produced a few blockbuster films with romantic elements such as the *Crocodile Dundee* series, the heart-warming *Strictly Ballroom* (1993) or the visually lush but saccharine Baz Luhrmann romantic epic, *Australia* (2008), many of our most iconic films – such as *Gallipoli* (1981), *The Adventures of Priscilla, Queen of the Desert* (1994), or *Muriel's Wedding* (1995) – are about homosocial friendships, or mateship, rather than the celebration of heterosexual romantic love. Indeed, Debi Enker observes that:

> Examining romantic relationships between men and women in Australian films is illuminating, if only as a guide to the filmmakers' collective discomfort with heterosexual love stories and scepticism about the possibility

of enduring passion. There is a striking absence of the grand passions that are intrinsic to, and characteristic of, French and American cinema . . . perhaps that is why Australian films have repeatedly examined bonds between men and shied away from love stories between men and women. Certainly there are very few films that depict a romantic relationship from the first flutter of attraction through to a blissfully happy ending for the couple. . . . Australian cinema seems sceptical about the capacity of love, and particularly passion, to endure. And even when it flickers for a while, it generally dies.

(Enker 1994: 218, 220)

Australian music seems similarly beleaguered by ambivalent representations of romantic love. Toby Martin (2017: 258) argues that 'Absence and loneliness were strong features of Australian hillbilly music and the sweetheart's absence served to illustrate the lonesomeness of the prairie and the loneliness of the singer'. While these songs are romantic, Martin argues that 'love was usually something sacred, idealised and distant' rather than about practices of intimacy (258). Meanwhile, in her study of romance in the lyrics of Australian pop and rock music, Michelle Arrow (2017: 292) suggests that unlike the heterosexual romantic sentimentality of Australian pop music, 'Oz Rock' was often more interested in articulating a 'discourse of sentiment for the abstract nation rather than romance for a single person' (297). The focus in the late twentieth century was on sexual desire, pleasure and loss. In the 'Indie' music trend of the 1990s, 'Romance was dead, and in its place was an approach to love songs that focused on irony, humour, and sentiment' (302). Arrow argues that the Australian discomfort with interpersonal romantic relations also manifests itself at the turn of the twenty-first century, when pop songs and dance music that apparently appear to be about love actually display 'a burgeoning lyrical focus on narcissism over romance, and of an emphasis on the independent self as more important than being part of an idealised romantic duo'. In these songs, a preference for 'casual, unattached, and unemotional sex was infinitely preferable to the complications of a relationship based on romantic love' (304–305). Arrow concludes that although romantic love has been an enduring theme in Australian music, where pop songs increasingly focus on the romance of the self, speaking 'of love for the self, rather than for a partner', Australian rock music tends to 'focus on non-romantic types of love: love gone wrong, homosocial love, love of nation, sentiment, and – increasing – love steeped in irony' (311).

Yet if positive affirmations of romantic love are hard to find in Australian culture, the fact remains that as a nation we still want to believe in an American narrative about romantic love that locates the state of living 'happily ever after' in marriage with our one true love. Despite the cogent feminist and queer critiques of marriage that were produced during the 1970s and 1980s (see Boucher and Reynolds 2017), despite the tomes of feminist scholarship showing just how awful heterosexual gender relations and the history of marriage has been, particularly in colonial Australia, access to same-sex marriage has been demanded as both a moral, romantic and citizenship right from 2004 until December 2017, when the federal parliament legalised same-sex marriage. In Australia, romantic love and intimacy have always encompassed much more than the romantic couple or their family or social circle. Questions of who was 'worthy' to be loved, who was considered 'marriageable' and who could be married have always been mixed up with political questions of the right to free and full citizenship from the colonial era right up until the present day.

Hsu-Ming Teo
Author of *Love and Vertigo*
(Winner of the Australian Vogel's Literary Award 1999)

Bibliography

Arrow, Michelle. 2017. "Love Is in the Air? Love in Australian Rock and Pop Music." In Hsu-Ming Teo (ed.), *The Popular Culture of Romantic Love in Australia*. Melbourne: Australian Scholarly Publishing, 285–318.

Arrowsmith, Robyn. 2013. *All the Way to the USA: Australian WWII War Brides*. Self-Published.

Bailey, Beth. 1988. *From Front Porch to Back Seat: Courtship in Twentieth Century America*. Baltimore: Johns Hopkins Press.

Bailey, Matthew. 2017. "'Ill-Natured Cartels of Anonymous Spite and Abuse': The Rise and Decline of Valentine's Day in Nineteenth-Century Australia." In Hsu-Ming Teo (ed.), *The Popular Culture of Romantic Love in Australia*. Melbourne: Australian Scholarly Publishing, 65–90.

Behrendt, Larissa. 2000. "Consent in a (Neo)Colonial Society: Aboriginal Women as Sexual and Legal 'Other'." *Australian Feminist Studies*, 15(33): 353–367.

Bongiorno, Frank. 2017. "Sex, Love and Romance in Twentieth-Century Australia." In Hsu-Ming Teo (ed.), *The Popular Culture of Romantic Love in Australia*. Melbourne: Australian Scholarly Publishing, 319–340.

Boucher, Leigh, and Reynolds, Robert. 2017. "Same-Sex Love in Late Modern Australia: On the Political Straight and Narrow?" In Hsu-Ming Teo (ed.), *The Popular Culture of Romantic Love in Australia*. Melbourne: Australian Scholarly Publishing, 341–368.

Cherlin, Andrew J. 2010. *The Marriage-Go-Round: The State of Marriage and the Family in America Today*. New York: Vintage.

Collins, Marcus. 2003. *Modern Love: An Intimate History of Men and Women in Twentieth Century Britain*. London: Atlantic Books.

Conor, Liz. 2004. *The Spectacular Modern Woman*. Bloomington: Indiana University Press.

Damousi, Joy. 1997. *Depraved and Disorderly: Female Convicts, Gender and Sexuality in Colonial Australia*. Cambridge: Cambridge University Press.

Dixon, Miriam. 1976. *The Real Matilda: Women and Identity in Australia, 1788–1975*. Blackburn, VIC: Penguin Books.

Dobrez, Patricia. 1983. "'The Human Form Divine': Martin Boyd in Love." In Axel Clark, John Fletcher and Robin Marsden (eds.), *The Theme of Love in Australian Writing: Colloquium Papers*. Sydney: Christopher Brennan Society, 3–16.

Elder, Catriona. 2017. "Romance and History on Australian Television." In Hsu-Ming Teo (ed.), *The Popular Culture of Romantic Love in Australia*. Melbourne: Australian Scholarly Publishing, 120–144.

Enker, Debi. 1994. "Australia and the Australians." In Scot Murray (ed.), *Australian Cinema*. St Leonards: Allen and Unwin, 210–225.

Finch, Lyn. 1995. "Consuming Passions: Romance and Consumerism During World War II." In Joy Damousi and Marilyn Lake (eds.), *Gender and War: Australians at War in the Twentieth Century*. Melbourne: Cambridge University Press, 105–116.

Flesch, Juliet. 2004. *From Australia with Love: A History of Modern Australian Popular Romance Novels*. Fremantle: Curtin University Press.

Green, Dorothy. 1983. "Love and the Thirteenth Chapter of Corinthians." In Axel Clark, John Fletcher and Robin Marsden (eds.), *The Theme of Love in Australian Writing: Colloquium Papers*. Sydney: Christopher Brennan Society, 43–56.

Greer, Germaine. 1993. *The Female Eunuch* (1970). London: HarperCollins.

Grimshaw, Patricia, Lake, Marilyn, McGrath, Ann, and Quartly, Marian. 1996. *Creating a Nation*. Ringwood, VIC: Penguin Books.

Grossi, Renata. 2014. *Looking for Love in the Legal Discourse of Marriage*. Canberra: ANU E-Press.

Grossi, Renata. 2015. "Understanding Law and Emotion." *Emotion Review*, 7(1): 55–60.

Illouz, Eva. 1997. *Consuming the Romantic Utopia: Love and the Cultural Contradictions of Capitalism*. Berkeley, CA: University of California Press.

Jamieson, Lynn. 2011. "Intimacy as a Concept: Explaining Social Change in the Context of Globalisation or Another Form of Ethnocentricism?" *Sociological Research Online*, 16(4). www.socresonline.org.uk/16/4/15.html

King, Andrew. 2017. "A History of Indigenous Marriage in Australia." In Hsu-Ming Teo (ed.), *The Popular Culture of Romantic Love in Australia*. Melbourne: Australian Scholarly Publishing, 91–119.

Lake, Marilyn. 1990. "Female Desires: The Meaning of World War II." *Australian Historical Studies*, 24(5): 267–284.

Lake, Marilyn. 1992. "The Desire for a Yank: Sexual Relations between Australian Women and American Servicemen during World War II." *Journal of the History of Sexuality*, 2(4): 621–633.

Langhamer, Claire. 2013. *The English in Love: The Intimate Story of an Emotional Revolution*. Oxford: Oxford University Press.

Lystra, Karen. 1989. *Searching the Heart: Women, Men and Romantic Love in Nineteenth-Century America*. New York: Oxford University Press.

Maroske, Sara. 1985 "Evangelical Love Letters." *Melbourne Historical Journal*, 17: 18–26.

Martin, Toby. 2017. "The Prairie Is a Lonesome Place at Night: The Absent Sweethearts of Hillbilly Music." In Hsu-Ming Teo (ed.), *The Popular Culture of Romantic Love in Australia*. Melbourne: Australian Scholarly Publishing, 257–284.

McGrath, Ann. 1984. "'Black Velvet': Aboriginal Women and Their Relations with White Men in the Northern Territory, 1910–1940." In Kay Daniels (ed.), *So Much Hard Work: Women and Prostitution in Australian History*. Sydney: Fontana and Collins, 233–297.

Roach, Catherine. 2016. *Happily Ever After: The Romance Story in Popular Culture*. Bloomington: Indiana University Press.

Rothman, Ellen K. 1987. *Hands and Hearts: A History of Courtship in America*. Cambridge, MA: Harvard University Press.

Russell, Penelope. 2017. "Love in a Colonial Climate." In Hsu-Ming Teo (ed.), *The Popular Culture of Romantic Love in Australia*. Melbourne: Australian Scholarly Publishing, 39–64.

Russell, Penny. 1988. "For *Better* and for *Worse*: Love, Power and Sexuality and Upper-Class Marriages in Melbourne, 1960–1880." *Australian Feminist Studies*, 3(7–8): 11–26.

Seidman, Steven. 1991. *Romantic Longings: Love in America, 1830–1980*. New York: Routledge.

Shumway, David R. 2003. *Modern Love: Romance, Intimacy and the Marriage Crisis*. New York: New York University Press.

Simmonds, Alecia. 2005. "'Promises and Pie-Crusts Were Made to be Broke': Breach of Promise or Marriage and the Regulation of Courtship in Early Colonial Australia." *The Australian Feminist Law Journal*, 23: 99–120.

Stone, Lawrence. 1988. "Passionate Attachments in the West in Historical Perspective." In Willard Gaylin and Ethel Person (eds.), *Passionate Attachments: Thinking About Love*. New York: Free Press, 15–27.

Summers, Anne. 1975. *Damned Whores and God's Police: The Colonization of Women in Australia*. Blackburn, VIC: Penguin Books.

Teo, Hsu-Ming. 2004. "The Britishness of Australian Popular Fiction." In Kate Darian-Smith, Patricia Grimshaw, Kiera Lindsey and Stuart Macintyre (eds.), *Exploring the British World*. Melbourne: RMIT Publishing, 721–747.

Teo, Hsu-Ming. 2005. "Love Writes: Gender and Romantic Love in Australian Love Letters, 1860–1960." *Australian Feminist Studies*, 20(48): 343–361.

Teo, Hsu-Ming. 2006. "The Americanisation of Romantic Love in Australia." In Ann Curthoys and Marilyn Lake (eds.), *Connected Worlds: History in Transnational Perspective*. Canberra: ANU E-Press, 171–192.

Teo, Hsu-Ming. 2014. "'We Have to Learn to Love Imperially': Love in Late Colonial and Federation Australian Romance Novels." *Journal of Popular Romance Studies*, 4(2). http://jprstudies.org/2014/10/we-have-to-learn-to-love-imperially-love-in-late-colonial-and-federation-australian-romance-novelsby-hsu-ming-teo/

Teo, Hsu-Ming, ed. 2017. *The Popular Culture of Romantic Love in Australia*. Melbourne: Australian Scholarly Publishing.

Weisser, Susan Ostrov. 2013. *The Glass Slipper: Women and Love Stories*. New Brunswick, NJ: Rutgers University Press.

Zwicky, Fay. 1983. "Speeches and Silences." In Axel Clark, John Fletcher and Robin Marsden (eds.), *The Theme of Love in Australian Writing: Colloquium Papers*. Sydney: Christopher Brennan Society, 29–41.

PREFACE

This book grew out of a course entitled 'Love and Intimacy at Contemporary Society' which I taught at a university in the UK. This proved to be incredibly popular with students, and I loved teaching it. The enthusiasm of these students towards the course between 2015 and 2017 encouraged me to develop the book which will serve as a key source of reading when the course is delivered in the future. Previously I published the book *Genealogies of Emotions, Intimacy and Desire: Theories of Changes in Emotional Regimes from Medieval Society to Late Modernity* (Routledge New York, 2017) with Routledge, which was a more theoretical research monograph and explored the work of a range of major theorists. This book is more focused on contemporary research on love and intimacy; it works to engage with the lives of those who experience love and intimacy, not just academics. At the same time, the book shows how relationships have changed significantly with the intervention of social media and the growth of dating websites, as well as the powerful #MeToo Movement. As I conclude this book, I am delighted to say that Sanne Mestrom, the wonderful artist whose work adorns the front cover of the book, has given permission for her work *Soft Kiss* to appear as the front cover of the book. This work is displayed in the Museum of Contemporary Art in Sydney, Australia, where I am completing this book.

Ann Brooks,
Sydney 2018

ACKNOWLEDGEMENTS

I wish to acknowledge the support of a number of people and institutions in the publication of this book. My thanks to Professor Bryan Turner, Director of the Institute of Religion, Politics and Society at Australian Catholic University, and to Australian Catholic University in Sydney and Melbourne for the invitation to be a Visiting Professor in 2018–2020. I completed the book in Sydney in an intellectually vibrant environment. My deepest thanks to artist Sanne Mestrom for allowing me to draw on her stunning work *Soft Kiss* for the front cover of this book. My thanks to Ursula and Robert of the Sullivan and Smith Gallery, who represent Sanne Mestrom, for their permission. My thanks also to Emerald Dunn Frost, Curatorial and Digital Administrative Assistant at The Museum of Contemporary Art in Sydney (who display this remarkable artwork), for supporting our efforts to draw on the artwork. My deepest thanks to the Director of Routledge, and to Emily Briggs and Elena Chiu at Routledge, for supporting the funding of this wonderful artwork for the front cover of this book. I wish to acknowledge the support of the Australia Research Council's Centre of Excellence for the History of Emotions during the period 2011–2018, who supported the development of this book by providing funding for me to undertake research for this book at the Huntington Library in Pasadena, California. My particular thanks to Katrina Tap and Tanya Tuffey in facilitating this. My thanks to students at Bournemouth University, who provided enthusiasm and support in undertaking the course 'Love and Intimacy in Contemporary Society' between 2015–17. My thanks to University of Sydney's Mandelbaum House, who created a marvellously vibrant location to work on the book. Finally, my deepest thanks to Emily Briggs, Sociology Editor at Routledge, who has been a remarkable source of support and fun in bringing this book to production, and who facilitated Routledge's support of the artwork *Soft Kiss* by Sanne Mestrom; and to Elena Chiu, Senior Editorial Assistant for Sociology, for her amazing efficiency.

Ann Brooks
Sydney 2018

INTRODUCTION

Background

The book *Love and Intimacy in Contemporary Society: Love in an International Context* reflects on relationships in contemporary society and the role of love and intimacy in framing lives. It looks at how love and intimacy is experienced differently and intersected by gender, ethnicity, race and sexuality. The book draws on sociological perspectives, cultural sociology and gender theory perspectives. The book aims to encourage people to understand theories of intimacy, emotions and desire by examining these concepts contemporaneously and cross-culturally. The book examines how love and intimacy is experienced by young people and how it is impacted by age. It looks at the representation of love and intimacy in the media and film and focuses on how gender, ethnicity and sexuality offer different perspectives on love and intimacy. The book shows how relationships are impacted by social networking and new technologies and the opportunities and challenges posed by these new platforms for building relationships. Finally, the book examines how intimacy has become commercialised in late capitalism and how that acts to change relationships. The book includes chapters on: love in a globalised world; diversity in love and intimacy in contemporary society; being young and in love; love in the movies; love and intimacy in marriage; modern romance; adultery, love and social networking; new technology, intimacy and work; and the commercialisation of intimacy. The book is written in an accessible way and explores a range of theoretical debates and contemporary research around emotions.

Perspectives on love and intimacy

It is argued that the sociology of love is an under-researched area and the reason, as Toye (2010: 4) observes, could be related to the association of love with 'the realm of women, the home, the private, the apolitical, the "not serious", that struggling to be taken seriously, feminist theorists feel such a "nervousness around the topic"'. However, this is not an entirely accurate assessment, as feminist and critical theorists have long engaged with the study of and implications of love.

Second-wave feminists such as Shulamith Firestone (1970), Simone De Beauvoir (1972) and later theorists such as Mary Evans (1998) were highly critical of romantic love and its implications for women. Firestone saw romantic love as the pivot of oppression for women. Both Firestone and de Beauvoir maintained that romantic love reinforced inequalities of power between women and men. De Beauvoir (1972: 669) was unequivocal in her criticism of romantic love, she viewed it as: 'a curse that lies heavily upon a woman confined in the feminine universe, woman mutilated, insufficient unto herself. The innumerable martyrs to romantic love bear witness against the injustice of a fate that offers a sterile hell as ultimate salvation'.

Mary Evans (1998: 273) has been consistently dismissive of romantic love in terms of the impact on women and men, but particularly women. She sees love as being ideologically corrosive: 'accumulated evidence of the last centuries suggest that people in the West have suffered more in their personal lives from "love" than any other single ideology'. Evans goes further and links romantic love with rape and violence against women.

Many feminist theorists see love as operating within the structures of patriarchy which by definition contribute to: 'the subordination of women (De Beauvoir, 1972; Dempsey, 2002; Ferguson, 1989, 1991, 2012; Firestone, 1970; Gunnarson, 2011, 2014a, 2014b; Illouz, 2012; Jackson, 2014; . . .) and privileging heterosexual love over same-sex love (Butler, 1990; Ferguson, 1989, 1991; Jackson, 2006)'(Gunnarson et al., 2018: 3).

Moving beyond discussions of patriarchy, Eva Illouz (2012: 112) as one of the contemporary feminist theorists of love, sees love as central in conceptualisations of individualism as she comments:

> . . . the sense of self-worth provided by love in modern relationships is of particular and acute importance, precisely because at stake in contemporary individualism is the difficulty to establish one's self-worth and because the pressure for self -differentiation and developing a sense of uniqueness has considerably increased with modernity.

Other feminists and queer theorists have focused on the relationship of romantic love and same-sex marriage. Grossi (2018) shows the problematic relationship of marriage for the queer community as shown in the work of a number of theorists. Judith Butler (2002) sees a discrepancy between the idea of marriage and those who espouse a radical sexual culture. As Butler (2002: 21) comments: '[f]or a progressive sexual movement, even one that may want to produce marriage as an option for non-heterosexuals, the proposition that marriage should become the only way to sanction or legitimate sexuality is unacceptably conservative'.

The clear view of theorists such as Berlant and Warner (2000) is that 'same-sex marriage replicates and privileges heterosexuality and further works to stigmatise other relationships' (Grossi, 2018: 8). The issue for the LGBT (lesbian, gay, bisexual and transsexual) community is: 'how to reconcile their opposition to marriage as a fundamentally conservative institution with a desire to support a campaign for recognition of LGBT relationships, and equality' (Grossi, 2018: 8).

A number of queer theorists (Berlant and Warner, 2000) have maintained that romantic love is inextricably tied to heterosexuality. However, Berlant, as one of the most innovative theorists in the field, argues that romantic love can be a site of optimism, change and transformation. Berlant and Warner (2000: 448) state that: 'love approximates a space to which people can return, becoming as different as they can be from themselves without being traumatically shattered, it is a scene of optimism for change, for a transformational environment'.

The basis of the optimism around romantic love for Berlant is its ability to accommodate difference, as she comments in relation to how we respond to romantic love: we cannot be seen as being all alike, but must understand our difference. As Berlant and Warner (2000: 448) show, '[we cannot assume] that we are all alike and compelled to repeat our alikeness intelligibly, but by teaching some of what we've learnt about love, under the surface, across the lines, around the scenes, informally'.

For Berlant, the 'queering of romantic love' is about its operation outside of established institutions. However, Eva Illouz maintains that the operation of romantic love in social structures is about the operation of power, with men defining the terms of the debate as Illouz (2012: 6) comments: '[Love is played out] in the marketplace of unequal competing actors where some people, mostly men, are able to command greater capacity to define the terms in which they are loved by others'.

So where do men stand in relation to love and emotions currently? In an interesting article by de Boise and Hearn (2017) entitled 'Are Men Getting More Emotional? Critical Sociological Perspectives on Men, Masculinities and Emotions', they argue that men's engagement with love and emotions has changed and they explore the implications. They show that recent research has challenged the idea that men are 'emotionally inexpressive'.

De Boise and Hearn (2017: 780) argue that current research shows that 'men not only have an active understanding of their emotional lives (see e.g. Galasinski, 2004), but in many cases appear to practice a "more emotional" form of masculinity (Forrest, 2010; Holmes, 2015; Roberts, 2013) than previously documented or assumed' (de Boise and Hearn, 2017: 780).

The development of research on men, love and emotions has been informed by engagement with both feminists and critical research and studies. De Boise and Hearn (2017: 781) maintain that 'cross-national, comparative studies demonstrated that gender affects emotional displays, which differ from country to country (Fisher and Manstead, 2000) . . . This suggests that displays of emotionality are conditioned by cultural factors which can be contested'.

Men have taken a long time to develop self-reflexivity, and that is clear in classical sociological theorising. De Boise and Hearn (2017: 782) note that 'Durkheim, Weber, Goffman, Elias and Le Bon have implicitly studied how men's emotions are connected to social structures (see Barbalet, 2002; Hughes, 2010) even if they often did not always recognize the gendered dynamic' (see also Brooks, forthcoming 2020).

However de Boise and Hearn (2017: 783) show that more recent social science research has given more attention to men's emotionality: 'in terms of both men and doing emotion work in (heterosexual) relationships than previously assumed (Holmes, 2015; Roberts, 2013) and not being afraid, to express relationships (Allen, 207; Forrest, 2010)'.

Perhaps of greater interest is how this more recent research challenges traditional theoretical modes of men and emotions, as illuminated by de Boise and Hearn (2017: 283):

> Much of this growing body of literature has demonstrated that men both have an understanding of their own emotional lives (Galasinski, 2004) and are more prepared to show emotions in front of other men (Anderson, 2000), leading to a 'softening'(Roberts, 2013) of, or a challenge to, 'hegemonic' masculinity/ies (Lomas et al., 2016; Montes, 2013).

Aims and objectives

The main aims, theories and objectives of the book are as follows: firstly, to understand contemporary thinking about love, intimacy and emotions and how they impact relationships in society; secondly, to develop an awareness of theoretical debates in sociology, cultural theory and gender theory, and how they resonate with everyday lived experiences of love, intimacy and emotion; and thirdly, to critique and analyse these debates in order to understand some key concepts and models for understanding the social world. These concepts include the transformation of intimacy; globalization and intimacy; 'non-standard intimacy'; online dating; 'love at a distance'; 'mail-order brides'; 'relational precarity'; 'spouse-busting'; emotional labour, reflexivity and modernity; and representation of love and intimacy in television and filmic and media theory. Finally, and fourthly, this book aims to understand the role of social networking and new technology in providing new ways of understanding how relationships are formed.

The contents of the book are varied, and cover theoretical perspectives, empirical research and case studies. Chapters include the following:

Chapter 1. Love in a globalised world
Chapter 2. Diversity and intimacy in contemporary society
Chapter 3. Being young and in love

Chapter 4. Modern romance

Chapter 5. Love in the movies

Chapter 6. Love and intimacy in marriage

Chapter 7. Adultery, love and social networking

Chapter 8. New technology, intimacy and work

Chapter 9. The commercialisation of intimacy

A synopsis of each of the chapters follows.

Chapter 1: Love in a globalised world

Chapter 1 focuses on how love and intimacy has changed in late modernity, as a result of globalisation. It considers how relationships of love and intimacy differ in different societies as a result of changing lifestyles.

The chapter provides a brief historical overview of how love and intimacy have been constructed in different historical periods, and also how love has been represented by sociologists, writers and feminist scholars. It shows how sociological interest in love and intimacy was much later than in literature and history. The chapter reflects on how literature, and in particular women writers, represent love and intimacy. It also looks at how some classical theorists viewed love in their work, and considers some of the more recent sociological interest in love and intimacy from theorists such as Luhman (1986), Giddens (1992), Bauman (2003), Beck and Beck-Gernsheim (1995, 2014) and Illouz (2012). This chapter examines how different theorists (Giddens, 1992; Shumway, 2003) have reflected changes in the nature of love and intimacy. This chapter also shows how the focus of sociology differs from that of psychotherapists and psychoanalysts.

The chapter also looks at contemporary perspectives on love and marriage through considering celebrity relationships and coupling, and the kind of images they convey, as well as on the royal marriages – in particular, the Royal Marriage of Prince Harry to an American actress, and how this royal marriage to an American differs so radically from the earlier marriage of a Windsor to an American. A comparison between the marriage Wallis Simpson and the Prince of Wales and the marriage of Harry and Meghan provides a yardstick of how attitudes to relationships have changed even among the royals.

Chapter 2: Diversity and intimacy in contemporary society

Chapter 2 looks at the relationship between intimacy and diversity, and considers explanations for this. Jamieson (2011) focuses on social change in the context of globalisation, while Roseneil and Budgeon (2004) discuss cultures of intimacy, with both sets of theorists focusing on intimacy in the context of the family. This chapter focuses on a range of issues covering heteronormative relationships and beyond, including 'non-standard intimacies' (Berlant and

Warner, 2000); 'the transformation of intimacy'; intimacy and emotion work in same-sex relationships (Roseneil, 2000; Umberson et al., 2015); and intimacy, sex and boundaries within intimate relationships (Elliott and Umberson, 2008). The chapter reflects on diversity in feminist perspectives on love, and also considers how the media have taken the lead in representing diversity in relationships. The chapter also shows how feminist and queer theorists have presented contested views on love and analysed love as both liberating and oppressive. It also focuses on how love can be an important dimension of the same-sex marriage debate.

Chapter 3: Being young and in love

Chapter 3 looks at how young women, and young people more generally, exercise choice in their lives as a result of education and occupational opportunities, and how this impacts their intimate relationships. The chapter draws on a wide range of contemporary studies, including research undertaken by the Pew Research Center (2014) in the United States, who investigate the relationship between teens, technology and romantic relationships. They examine the role that social and digital media play in romantic relationships of youth in America. A wide range of sources show how young people are negotiating romance and intimacy. Willmot (2007) explores the relationship between young women, opportunities through education and employment and how they construct love and intimacy. Richtel (2012) looks at how new technology plays into relationships with younger people and their response to it. Elley (2015), in a fascinating study of 'laddishness' in higher education, looks at the growth of inappropriate, abusive and misogynistic behaviour in universities. Elley considers the implications of this for gender politics in higher education and the growth of misogynistic behaviour. The chapter also looks at the work of Bogle (2008) on how relationships for young people have moved from dating to 'hooking up', which redefines the nature of love and intimacy in relationships.

The chapter also looks at the historical impact of romance comic books on love and romance among young people in Australia, and how many of the stories and images were adapted from an American context to suit an Australian setting.

Chapter 4: Modern romance

Chapter 4 focuses on the humorous exploration of contemporary romance by the comedian Aziz Ansari (2015b) in his book *Modern Romance*. The book was written with the assistance of the sociologist Eric Klinenberg, and offers an interesting exploration of how people find love in contemporary society. Issues include meeting partners, the role of technology, and sexting, cheating and breaking up. A wide range of issues are raised by Ansari which relate to intimacy and marriage, including the following: geographical proximity and partnership; average age of first marriage; choices in marriage patterns; online dating; how people met future partners; adultery and internet cheating; passionate and companionate love; and marriage and commitment.

The Americanisation of love and intimacy is very evident in both sociological and psych-social research on relationships as well as the commercialisation of intimacy. This chapter reviews some of the influences on the commercialisation of love and intimacy in an Australian context. It also finally reflects on whether chick lit and romance novels offer new versions of romance, or confirm traditional narratives of romance through the commercialisation of intimacy.

Chapter 5. Love in the movies

This chapter focuses on how love and intimacy has been represented in films. We look at a range of films, from the United States and elsewhere, to make comparisons about how relationships are represented in different cultural contexts. The chapter look at the way romance has been represented historically with the emergence of different genres including film noir, romantic comedies and bromances ('homme-coms').

The chapter also explores the way in which the '#MeToo Movement' has impacted how women are represented in all aspects of the production of movies and the way in which it has undermined powerful individuals in the film and media industries, including Harvey Weinstein and Les Moonves, through the significant impact of the reporting of Ronan Farrow in *The New Yorker*. The impact continues to be significant on women in the television and filmic industries.

The chapter also looks at heteronormative perspectives on love in the movies and examines movies where challenges are made to the heteronormative dominance of movies. The chapter looks at the psychosocial context of jealousy and socio-cultural construction of masculinity and jealousy. Yates (2007) explores the socio-cultural construction of jealousy and how it is represented in film. Stearns (1989) tracks the history of jealousy in American history. Mullen (2018) argues that whereas jealousy was once an accepted public emotion, male jealousy was linked to a male sense of entitlement. Mullen (2010) also notes that historically this changed in the twentieth century with the emergence of feminism, and male jealousy is seen in a different way and linked to domestic violence and stalking. This chapter explores a number of cinematic theories which examine a psychosocial approach to understanding emotions in film including Creed (1993) and Mulvey (1975, Mulvey et al. 2015).

Chapter 6: Love and intimacy in marriage

Chapter 6 explores historical and cultural perspectives on marriage and traces different forms of marriage. Coontz (2016) highlights the fact that traditionally, marriage benefited men rather than women. Additionally Coontz (2015) looks at marriage and social change in the United States. This chapter also explores the emergence of same-sex marriage in Australia and the United States.

Sociologists such as Beck and Beck-Gernsheim (2014) have shown how individualism which has accompanied a romanticising idea of absolute love has undermined the traditional forms of living together. The classic family consisting of a man, woman and children has given way to a multiplicity of new types of family. The Beck and Beck-Gernsheim (2014) argue that husbands are replaced by 'serial' partners, single mothers and fathers are more the norm, as are 'patchwork' families and relationships, a product of serial marriages and divorces.

An additional area explored in this chapter is the area of migration, marriage and intimacy through the work of Beck and Beck-Gernsheim (2014) which looks at 'commercial matchmaking', and mixed-nationality marriages as well as women who migrate in search of marriage. Beck and Beck-Gernsheim 2014: 2 describe 'love at a distance' and the 'the global chaos of love', with mixed-nationality couples, people who migrate for work or marriage, women who rent out their wombs and Skype-based love relationships as examples of this. The work of Beck and Beck-Gernsheim (2014) provides an important contribution to how relationships and families have changed.

Constable (2005a, 2005b, 2006) has written extensively on the role of marriage and migration and explores what she calls the 'cultural logic of desire'. The chapter also examines the issue of 'mail-order brides'. This chapter looks at heterosexual and same-sex marriage and considers why marriage is still an important factor in relationships and how attitudes towards what constitutes marriage and family has changed.

Chapter 7: Adultery, love and social networking

Chapter 7 explores adultery in relationships and in its representation in the media. We examine the significance of adultery in the twenty-first century and assess whether it carries the same moral significance as it has in the past (Leonard, 2010a). We are particularly interested in the role gender plays in adultery and whether the role of women in relationships has fundamentally shifted. Leonard (2010a) examines the relationship between adultery tropes and working women. She also explores how marriage is central in women's lives, and shows how this can be seen in the multi-billion dollar wedding industry. This chapter explores this in the 'celebrity marriage', and looks at the Clooney marriage. It also looks at the HBO original series *Sex and the City* and how it combines glamour, fashion and celebrity in marriage. Leonard shows how adultery is represented in a number of high-profile Hollywood films. The chapter explores a range of research both classical (Kipnis, 2003) and contemporary (Buunk and Dijkstra, 2006) to assess the historical and contemporary significance of infidelity (Fisher et al., 2010).

The chapter explores the historical context of adultery (Turner, 2002) and shows the significant gendered discourses surrounding adultery (Gregg, 2013). We also examine a more contemporary analysis through the concept of 'spouse-busting'. The chapter also examines theoretical contributions from Berlant (2011), who explores relationships and precarity (Brooks, 2017). The chapter explores the case of *Ashley Madison* in the United States, which casts itself as the website for establishing adulterous relationships. The chapter also explores the hack of the *Ashley Madison* website and its implications.

Chapter 8: New technology, intimacy and work

This chapter looks at how new media technology and its effect on intimacy impacts everyday life and how it influences the work-life balance. This chapter looks at Facebook and other social networking platforms to assess what impact this has on everyday life and intimacies. Gregg (2011) considers the relationship between new media technologies and intimacy in relation to people's homes. This chapter also examines the relationship between technology, intimacy (Francisco, 2013) and transnational families. The chapter also examines other cultural contexts in relation to intimacy and technology (Hannaford, 2015) which includes 'intimate surveillance'. Additionally Parrenas (2005b) raises the issue of 'long distance intimacy' and, elsewhere (Parrenas, 2014), examines the constitution of intimacy in the use of communication technology in Filipino transnational families.

Chapter 9: The commercialisation of intimacy

This chapter examines the relationship between romance and consumption as outlined in the work of Eva Illouz, and also explores postmodern love, which Illouz sees as captured in Candace Bushnell's *Sex and the City*. The chapter thus focuses on the relationship between *Sex and the City*, consumer culture and post-feminism as an example of one aspect of the commercialisations of intimacy and considers what a postfeminist woman-centred drama amounts to. The chapter also explores the relationship between postmodern romance and the emergence of the love affair, it also considers the changing nature of love affairs in gendered terms. The chapter finally looks at the sexualisation of modern relationships. This chapter examines feminist theorists' analysis of the commercialisation of intimacy. Illouz's *Hard Core Romance: Fifty Shades of Grey* examines a range of controversial and popular literature which focuses on intimacy.

Conclusion

Throughout the pages of this book, I show how love and intimacy is an academic question, but also a very personal issue which has implications for all aspects of social life. A well-documented and clearly defined sociology of emotions to include love and intimacy is long overdue.

1

LOVE IN A GLOBALISED WORLD

Love looks not with eyes, but with the mind
And therefore is winged Cupid painted blind.
Nor hath love's mind of any judgement taste,
Wings and no eyes figure unheedy haste.
And therefore is love said to be a child
Because in choice he is so oft beguiled.
　　　　　Shakespeare, W. A Midsummer
　　　　　　　　　Nights Dream

Romantic love accompanied the dawning of individualisation among the bourgeoisie, and it only became a dominant discourse once individualisation was widespread.
　　　　　Shumway, D. (2003) Modern Love: Romance, Intimacy and the
　　　　　Marriage Crisis *(New York: New York University Press)*

Introduction

This chapter provides a brief historical overview of how love and intimacy have been constructed in different historical periods, and also how love has been represented by sociologists, writers and feminist scholars. It shows how sociological interest in love and intimacy arose much later than in literature and history. It looks at how some classical theorists viewed love in their work, and considers some of the more recent sociological interest in love and intimacy from theorists such as Luhman (1986), Giddens (1992), Bauman (2003), Beck and Beck-Gernsheim (1995, 2014) and Illouz (2012). This chapter examines how different theorists (Giddens, 1992; Shumway, 2003) have reflected changes in the nature of love and intimacy.

The chapter also looks at contemporary perspectives on love and marriage through considering celebrity relationships and coupling, and the kind of images they convey, as well as on the royal marriages – in particular the Royal Marriage of Prince Harry to an American actress, and how this royal marriage to an American differs so radically to the earlier marriage of a Windsor to an American. A comparison between the marriage of Wallis Simpson and the Prince of Wales and the marriage of Harry and Meghan provides a yardstick of how attitudes to relationships have changed even among the royals.

Romantic love as a social construction: modern and pre-modern love

In Brooks (2017), I showed how romantic love can be viewed as a social construction which has changed over time and with different historical periods. In this book, we are concerned with providing an analysis of love and intimacy in contemporary society, but it is useful to reflect from a sociological perspective on the changing nature of romance, love and intimacy.

As I indicated previously (Brooks, 2017), the history of love and intimacy shows a fascinating interweaving of social, economic and cultural influences. Shumway (2003: 12) shows that: 'Romantic love . . . is best understood as a culturally specific discourse', which is most clearly defined in the history of Western culture.

In fact the traditional meaning of love does not relate directly to passion or intimacy:

> The traditional meaning of *love* is not romance but social solidarity: it corresponds to the capacity for bonding rather than the capacity for infatuation. For most of Western history, as Luhmann argues: 'What was considered important is not living out one's passions, but rather a voluntarily (but not completely or slavishly) developed solidarity with a given order' (Luhman, 1986: 130).
>
> *(Shumway, 2003: 12)*

Romance as a historical and cultural discourse has political and economic dimensions. In medieval society, romance emerged as an alternative discourse to conventional marriage which was officially sanctioned by the aristocracy. It offered a discourse which provided a cultural alternative, and offered shifts in manners and morals in the courts of 'feudal Europe' (Shumway, 2003: 13). In addition, there is also a focus on the idealisation of love. Shumway argues that part of this was the role that women played in idealising love. Previously, women had been seen as a corrupting influence, but they came to be idealised within the discourse of romance at the same time love was also idealised.

As we see below in the analysis of the classical theorists' contribution to understanding love, a clear distinction was drawn between passion, religion and sex. Giddens (1992) notes that the 'secular' use of the word 'passion' as distinct from 'religious passion' is a modern concept. Giddens shows when we think of 'passionate love' (*amour passion*) it implies a connection between love and sexual attachment. Giddens maintains that passionate love has never been seen as a basis for marriage, and in most cases is seen as disruptive. He correctly distinguishes between passionate love as a universal phenomenon and romantic love as culturally specific, which concurs with what Shumway (2003) argues.

Giddens (1992) notes that passionate love was directly related to class, and that marriages were contracted based on economic position rather than sexual attraction. He also notes that among the peasantry in seventeenth-century France and Germany, kissing, caressing and other forms of physical attraction were rare among married couples. However, it was common for men of all social classes to have affairs. Giddens notes that the aristocracy allowed 'respectable' women to have sexual liaisons. Women in the aristocracy were liberated from reproduction and routine work, which allowed them time to pursue sexual pleasures.

There is a vast literature on romantic love in the eleventh and twelfth centuries. Courtly love poetry was performed by poets who travelled from castle to castle performing for aristocratic women. As Swidler (2001: 112) notes, love was seen as an ennobling activity rather than a dangerous appetite. Lyrics told of knights who were 'made virtuous by love and of heroic deeds performed in the service of noble ladies'.

Swidler shows that courtly love remained the dominant code of the European nobility for centuries, but she says that the courtly ideal of love was gradually reshaped by the bourgeois culture of early capitalism (see below). Swidler (2001: 113) says that: 'rather than inspiring heroic deeds, love becomes a test of individual character . . . the bourgeois love story ends with a marriage in which the autonomous individual finds his or her proper place in the social world'.

Class and gender are all important points of intersection in the historical development of love and intimacy. David Shumway (2003: 7) shows that:

> Historians have long connected the rise of companionate marriage and the later association of romantic love and marriage to the rise of the bourgeoisie. The aristocrats needed marriages of alliance to preserve their power and wealth, and the working class typically married for their economic advantage that extra hands brought to the household.

It is the growth of the new middle class – the petit bourgeoisie and the professional middle class – who are positioned economically and educationally to develop new patterns of love and intimacy. The expansion of the professional managerial class throughout most of the twentieth century is one of the conditions that enabled the discourse of intimacy to develop.

Women became increasingly influential in the eighteenth and nineteenth centuries, particularly with the growth of romance novels, which were often written by them. Giddens (1992: 43) shows that the 'fusion' of romantic love and motherhood did give women a 'domain of intimacy'. For men, he says, there was a tension between romantic love and passionate love which resulted in a separation of feelings with the 'comfort of the domestic environment separated from the sexuality of the mistress or whore.' The double standard gave women no freedom from the domestic sphere.

As stated, one of the key drivers of popular and literary fiction was the growth of the print media. The impact of romantic fiction was both powerful and devastating. In his book *The Transformation of Intimacy: Sexuality, Love and Eroticism in Modern Society*, Giddens (1992) argues that romantic love introduced the idea of a narrative into an individual's life. Giddens argues that telling a story was one of the meanings of romance (see Brooks, 2017). The rise of romantic love more or less coincided with the emergence of the novel.

The romantic novel originated in Europe at the end of the eighteenth century and beginning of the nineteenth century and was located in a period which was characterised by romantic literature, poetry and art. The focus was on an intense 'courtship period', and the emphasis on emotion in the Romantic era moves away from an understanding of relationships being based on economic circumstances which had characterised relationships in pre-modern times.

Romantic love is seen by Evans (1998: 27) as a 'formative part of the gradual ideological emancipation of women and the public definition of a specifically feminine set of interests'. An important aspect of this was the exercise of choice in the selection of partners. Evans (1998: 27) makes the point that women, through the discourse of romantic love, could exercise choice in who they would marry and could construct male behaviour.

Evans shows there was a shift to 'men in love' reflected in the eighteenth century novel – heroes such as Darcy in Jane Austen's *Pride and Prejudice*, Tom Jones in the novel of the same name and Count Vronsky in Tolstoy's *Anna Karenina*. All the men in these novels became destabilised by their love for women. There was something of a backlash against the romantic novel, with Jane Austen's novels offering a humorous and scathing attack on love.

Jane Austen and the romantic novel: an early feminist critique

Austen made a major contribution as an English novelist to romantic fiction. Her work was popular and filled with satire and irony as she mocked the performance of romance within the British aristocracy and landed gentry. Her publications include *Sense and Sensibility* (1811), *Pride and Prejudice* (1813), *Mansfield Park* (1814) and *Emma* (1815). She also wrote *Northanger Abbey* and *Persuasion*, which were published after her death in 1818. Her plots were comic and highlight the dependence of women on marriage to secure social standing and economic security.

Austen's writings appeared historically during the period of British Romanticism, and Austen was sympathetic to a number of the Romantic poets, including William Wordsworth (1770–1850), Samuel Coleridge (1772–1834), John Keats and Byron, despite the misogynistic commentary and perspectives of many of the Romantics.

Austen had produced a range of earlier drafts and versions of her published work, but her first novel was *Sense and Sensibility*, published in 1811. *Pride and Prejudice* was published in 1813 and *Mansfield Park* in 1814. Mansfield Park was very popular, and all copies sold within six months. She moved publishers and published *Emma* in 1815, and a second edition of Mansfield Park in 1816. These novels were the last ones published in her lifetime. Her later novels were published by her family posthumously; *Persuasion* and *Northanger Abbey* were published in 1818. In 1832, the publisher Richard Bentley purchased the remaining copyrights to all of Austen's novels and published them in five illustrated volumes as part of his standard novel series. In October 1833, Bentley published the first collection of Austen's works.

Realist novels and erotic literature

Standing in comparison with the humour and satire of Jane Austen's novels is the realist novel and the romantic novel. An example of the realist novel is *Madame Bovary* by Gustave Flaubert (1856). Flaubert's novel reveals the motivations of women in unhappy relationships, and shows how women can initiate adulterous affairs and seek ways to escape from the boredom and emptiness of marriage. The novel was seen as erotic and was attacked for obscenity. It highlights how women can find married life dull and predictable, and motherhood disappointing. The novel's protagonist was attracted to relationships with younger men and has a series of affairs, which are described in detail. However, it ends with the heroine swallowing arsenic, as she is in debt, and dying. The interesting dimensions of the novel were Flaubert's realist style and social commentary. It also highlights the nature of social class in provincial France. The realist movement also was a reaction against romanticism.

Another novel which highlighted aspects of class in relationships was *Lady Chatterley's Lover* by D.H. Lawrence (1928). The novel focused on the physical aspects as well as the emotional aspects of relations between an upper-class woman and a lower-class man. An interesting dimension of the novel is the fact that it reflects on the fact that the female protagonist's husband is paralysed, and highlights the sexual frustrations of the heroine in her search for sexual gratification. It also reflects on the nature of heightened sexual experience, and deals with issues that are seen in other Lawrence novels such as, in *Women in Love*, the relationship of the mind and body. Lawrence reflects on the brutality as well as the romance of relationships.

Lawrence reflects the issue of class in Britain in the early twentieth century, and it is a recurrent theme in his books, including *Sons and Lovers* and *Women in Love*. When the book *Lady Chatterley's Lover* was published by Penguin Books in Britain in 1960 in its full unexpurgated edition, Penguin was tried under the Obscene Publications Act 1959. A verdict of 'not guilty' was given, and this made Lawrence's book available for the first time in Britain.

The undertheorisation of love in the sociological tradition

Rusu (2018) maintains that until recently, love was heavily undertheorised in the sociological tradition. He makes the point, perhaps unsurprisingly, that Karl Marx ignored love altogether in his work. Emile Durkheim's reflections on love are to a large extent concentrated in a single study in *L'Annee Sociologique*. He makes the distinction between family love and passionate love. Rusu (2018: 3) makes the point that: 'the two emotional types embody "the eternal antithesis between passion and duty" (Durkheim, 1897: 67). For Durkheim, family love stands under the mark of duty . . . as an expression of social morality. . . . Passionate love between two individuals, in contrast, is free love – *amour libre* – the outcome of "the movement of spontaneous private sensibilities" (Durkheim, 1897: 61).' Rusu points out that Durkheim had little else to say about love.

Rusu (2018: 5) goes on to show how conceptions of love were developed up to a point in the work of Max Weber, Sorokin and Talcott Parsons. Weber covered the concept of love in his work on the historical sociology of religion and the concept of 'brotherly love' in salvation religions. As Weber pointed out: 'the brotherly ethic of salvation religion is in profound tension with the greatest irrational force of life; sexual love' (Weber, 1946: 343). The reason for this is outlined by Rusu (2018: 10), who argues that:

> This is because passionate love for an erotic partner is exclusivist by default, and this feature of it undermines the universality and all-inclusivism of brotherly love. For this reason the theologians of salvation religions struggled to domesticate passionate love by restricting it to the confines of the marriage. Modernity brings this essential tension to fever pitch, as sexual activity has been sublimated into eroticism. The erotization of sexuality implies the transformation of 'the sober [sexual] naturalism of the peasant' into a refined sensual art of carnal love (Weber, 1946: 344). Through its careful and deliberate cultivation, eroticism had raised sexuality from the state of naïve naturalism into a 'sphere of conscious enjoyment'.
>
> *(Weber, 1946: 347)*

In other words, Rusu states that for Weber, erotic love is conceived as a means of redemption, of re-enchanting the world and as a means of offsetting the 'iron cage' of economic efficiency and instrumentalism. It is a promise of giving the modern individual a 'sensual religion of inner-worldly salvation' (Rusu, 2018: 11).

Perhaps the most ardent proponent of love was Sorokin. As Rusu notes: 'Sorokin launched an ambitious programme for studying love as an energetic force that has the transformative power to remould human personality and society alike' (Rusu, 2018: 12). In his book *Altruistic Love*, Sorokin (1950: v–vi) outlines the benefits of love in therapeutic terms:

> it will show that love is literally a life-giving force; . . . that love annuls loneliness and is the best antidote to suicidal morbid tendencies; that love experience is true cognition; that love-experience is beautiful and beautifies anything it touches; that love is goodness itself;. . . . that love is fearless and is the best remedy for any fear; that love is a most creative power; . . . that it is the best remedy against hate, insanity, misery, death and destruction.

Rusu notes that Sorokin's (1954) most 'audacious' attempt at theorising love as a 'transfigurative energetic force' is contained in his book *The Ways and Power of Love: Types, Factors, and Techniques of Moral Transformation*. In this, Sorokin links sociology with social psychology and his perspective is informed by Freud's psychoanalysis.

In *The Reconstruction of Humanity*, Sorokin (1948) sets out to show how a theory of love can work to assist mankind's 'self-destructive crisis'. Rusu (2018: 15) shows how:

> Sorokin laid the ground for a new science of love that he named 'amitology', characterised as an 'applied science of amity and unselfish love' (Sorokin, 1950). Amitology was designed by Sorokin as a science of social good, as an 'art of cultivation of amity, unselfish love, and mutual help in interindividual and intergroup relationships' (Sorokin, 1948: 277).

Sorokin in fact imagined a *political economy of love* which was based on 'finding and inventing the most efficient ways of production, accumulation and circulation of love energy in the human universe' (Sorokin, 1950: 278). Rusu points out that Sorokin, perhaps unsurprisingly, failed to convince the sociological community to follow his theory of amitology.

In fact, despite these efforts, Rusu points out that the classical theorists had relatively little to say about love. As Rusu (2018: 4) shows:

> The fascination with love was slow to gain traction in sociologists' hearts and minds, but it eventually burst out with the publication of works such as Niklas Luhmann's (1986) *Love as Passion: The Codification of Intimacy*, Anthony Giddens (1992) *The Transformation of Intimacy*, and Zygmunt Bauman's (2003) *Liquid Love: On the Frailty of Human Bonds*. These books were followed by other important contributions such as *The Normal Chaos of Love*, Beck and Beck-Gernsheim (1995) and *Why Love Hurts: A Sociological Explanation* by Eva Illouz (2012).

Modern love – romance, intimacy and the marriage crisis

The extension of individualisation, alongside the rise of capitalism, allowed the bourgeoisie to influence the direction of love and intimacy in countries such as Britain. However as Shumway (2003: 19) shows, the nineteenth century provided a wide range of responses to the relationship between love, intimacy and marriage, and he claims the relationship was more clearly defined in America than in Europe.

In her book *Love in America* (1987), Cancian shows that from the nineteenth century, gender roles became more polarised, with love being identified with emotional expression and being the responsibility of women, while self-development was regarded as a concern of men. Cancian (1987: 17) comments that in the early nineteenth century:

> marital intimacy, in the modern sense of emotional expression and verbal disclosure of personal experience were probably rare. Instead husband and wife were likely to share a more formal and wordless kind of love, based on duty, working together, mutual help and sex.

Cancian shows that by contrast, new images of love began to emerge, emphasising self-development for men and women. And more flexible and androgynous roles started to become acceptable from 1900, and accelerating in the 1960s.

More recently, sociologists such as Beck and Beck-Gernsheim (2014) have shown how individualism, which has accompanied a romanticising idea of absolute love, has undermined traditional forms of living together. The classical family consisting of a man, woman and children has given way to a multiplicity of new types of family. The development of more diverse relationships within and outside marriage is explored in this book (see Chapter 2). Beck and Beck-Gernsheim (2014) maintain that husbands are now replaced by 'serial' partners, and single mothers and fathers are more a norm, as are patchwork families, and relationships a product of successive marriages and divorces.

Celebrity love and marriage

Two examples highlight the increasing diversity of love, intimacy and marriage: first, celebrity marriages, second, royal marriages. One of the most high-profile celebrity marriages is that of George Clooney and Amal Alamuddin. In an article in *The Daily Telegraph* entitled 'Rise of the "Sapiosexual": Why A-List Men Date Brainy Women' by Karen Yossman (2018), Yossman says that George and Amal's marriage signalled a new attitude among Hollywood leading men. Amal is of course an A-list academic and hugely well qualified, as an Oxford-educated barrister and spokesperson on human rights cases. By comparison, George is a talented and outspoken political critic of Trump and the Republican Party. There is significant anticipation that he will run for political office in the United States, and as such to be linked with someone such as Amal would be regarded as a real asset.

The traditional pattern for Hollywood men has not been to prioritise intelligence and status when it comes to selection of partners, but the Clooney relationship has signalled a shift in the partnerships that have emerged more recently. Perhaps most notably this can be seen in Brad Pitt's recent partners. He is, of course, a close friend of the Clooneys. One of his recent partners is Neri Oxman, who is a 42-year-old professor at MIT and an award-winning artist whose work has been displayed in New York's Museum of Modern Art. Other famous pairings include Benedict Cumberbatch, who is married to an Oxford-educated theatre director wife, Sophie; actor Joseph Gordon-Levitt, married to Silicon Valley robotics expert Tasha McCauley; and Jude Law, married to Phillips Coan, a business psychologist with a PhD.

Yossman comments that on the dating site Plenty of Fish, more than 89,000 men identify as 'sapiophiles', while Match.com and OKCupid have added 'sapiosexual' as a category of orientation. There's even an entire dating app dedicated to them called *Sapio* which, according to its creator, is aimed at singles and focused more on the mind and the heart than simply on looks. Yossman (2018) comments that this development is a positive one:

> What is particularly encouraging about this courtship of alpha women is that their partners are content to bask in their glow, rather than feel emasculated. Certainly Clooney never seems to tire of waxing lyrical about Amal's achievements, which include representing Armenia at the European Court of Human Rights. He even makes jokes at his own expense, claiming in a recent *Vogue* profile: 'She's the professional, and I'm the amateur.'

Of course, the trilingual and highly educated Amal would certainly be able to be an outstanding First Lady if and when George decides to run.

As Yosser indicates: 'Beyond Hollywood, it looks like the rest of the world's male population is catching up. A U.S. study by Match.com revealed that 87 percent of men would date a woman who was their intellectual, academic and economic superior – a statistic that flies in the face of generations of preconceived wisdom that men fear intelligent women'.

The royals, love and marriage

An interesting view of the changing perceptions of global marriage within the British Royal Family is given by Andrew Morton (2018b) in his book *Wallis in Love: The Untold True Passion of the Duchess of Windsor*. In an article taken from the book entitled 'Meghan and Mrs. Simpson' in the *Telegraph Magazine* (2018a), Morton provides a fascinating comparison between the royals' attitude to the American divorcee Wallis Simpson, and their response to the American divorcee Meghan Markle. As Morton (2018a: 15) says: 'Two American divorcees marrying into the British Royal family: one was treated with disdain, the other with adulation. What a difference 80 years makes'.

Morton shows that Queen Mary considered Wallis a 'sorceress' for luring Edward VIII from his destiny and his duty. But the response of the royals to Meghan was entirely different, as Morton (2018a: 16) comments:

> How differently they were treated: former actress Meghan who divorced her film producer husband Trevor Engelson after less than two years of marriage, has been warmly welcomed into the royal bosom. She was invited to spend Xmas with the Royal Family at Sandringham even though she was not yet officially part of it, and she walked arm-in-arm with fiancé Harry, chatting with the Duke and Duchess of Cambridge after they left St Mary Magdalene Church after the Christmas Day service – her place firmly ahead of the Princess Royal, the Duke of York and Princesses Beatrice and Eugenie towards the front of that very hierarchical family procession. The duchess-in-waiting also received an oblique shout-out in the Queen's Christmas broadcast.

Morton points out that the treatment of Wallis Simpson could not have been more different, when Edward VIII abdicated from the throne, and both were in effect exiled from the realm. As Morton notes, for the rest of their lives they moved from one country to another, living in Paris, New York, the Bahamas and the South of France, despite the Duke offering to work to support the royal family. As Morton (2018a: 16) comments: 'Wallis Warfield Simpson was singled out as the primary culprit in the constitutional crisis that gripped the nation in the dying days of 1936'. In fact, Morton says that some senior government ministers thought she was a Nazi spy. Morton shows that: 'high-society gossiped that this rather manly looking woman had seduced the sovereign thanks to exotic sexual techniques that she had learnt in the sing-song houses of Hong Kong and Shangai'. Apparently, Wallis Simpson's response was 'Venom, venom, venom'. In fact, the issue of relationships and intimacy was interesting among the royals at that time, and Morton reports that when Edward, Prince of Wales met Wallis Simpson, it was at a house party of his mistress Viscountess Furness in 1931.

Meghan Markle, of course, is an American actress, but perhaps more interestingly for the royals, she is bi-racial and has an African-American mother and ancestors who worked as slaves on the cotton plantations in Georgia. As Morton comments: 'The bi-racial actress would probably not have been countenanced by the snobbish socialite Wallis Simpson.' In fact, both Meghan and Wallis had complicated family histories, and Wallis had been divorced twice.

Fashionistas, revenge and the royals

Morton shows how all three women who were married to royals but outsiders themselves, Wallis, Diana and Meghan, used fashion to express their feelings.

Morton refers to the 'revenge dress' worn by Diana to a charity event on the night that the Prince of Wales admitted his adultery on television. This was seen as the night that Diana expressed her independence. Wallis Simpson also used fashion to reflect her position:

> Wallis Simpson also used her wardrobe as a weapon, her sleek, crafted style in sharp contrast to the homely fashions preferred by her enemy, Queen Elizabeth, the Queen Mother whom she called 'Cookie' since she resembled a cook. While Wallis acknowledged that she was no great beauty, she made sure that her clothes reflected the standing of her husband as ex-king, preferring to wear such designers as Chanel, Givenchy or Dior.
>
> *(Morton, 2018a: 18)*

Meghan, as Morton shows, wears labels of ecologically and ethnically minded designers, as well as companies that have a philanthropic element in their business strategy. 'She once used her blog to promote such brands as Conscious Step (a sock company that plants 20 trees for every pair sold) and The Neshama Project (a jewellery business that donates a percentage of profits to Innovation Africa)'. In addition, on one of her recent visits, 'she carried a bag by De Mellier, a British label that funds life-saving vaccines through its sales, and a cruelty free coat by Stella McCartney. As she once noted in her blog: "It's good if you are fabulous but great if you do something of value to the world"' (Morton, 2018a: 19).

Conclusion

Chapter 1 shows how love and intimacy changes over time, and shows how sociologists have been slower than historians and literary writers in analysing the significance of love and intimacy in people's lives. The chapter reviews the perspectives of a range of sociological theorists on love and intimacy, including Giddens, Beck and Beck-Gernsheim, Swidler and Illouz. The chapter also looks at the role of the romantic novel and erotic literature in defining attitudes towards love and intimacy. The chapter looks at the undertheorisation of love in the sociological tradition, and considers the work of some of the classical theorists in this regard. Sociologists have also looked at changes in the nature of love and marriage. The final part of the chapter looks at the contemporary interest in celebrity love and marriage. The chapter concludes with an analysis of the Royal Marriage in the UK, where Prince Harry married the American actress Meghan Markle. A comparison is made between the earlier marriage of the Prince of Wales to Wallis Simpson, because she was a divorcee, and the change in social attitudes towards the Royals with the recent Royal marriage.

2

DIVERSITY AND INTIMACY IN CONTEMPORARY SOCIETY

Introduction

This chapter looks at the relationship between intimacy and diversity, considers explanations and explores a range of different feminist and modernist social perspectives. Jamieson (2011) focuses on social change in the context of globalisation, and considers the transformation of intimacy in considering diversity; Roseneil and Budgeon (2004) discuss cultures of intimacy, with both sets of theorists focusing on intimacy in the context of the family. This chapter focuses on a range of issues covering heteronormative relationships, including 'non-standard intimacies' (Berlant and Warner, 2000); 'the transformation of intimacy'; intimacy and emotion work in same-sex relationships (Roseneil, 2000; Umberson et al., 2015); and intimacy, sex and boundaries within intimate relationships (Elliott and Umberson, 2008). The chapter reflects on diversity in feminist perspectives on love, and also considers how the media have taken the lead in representing diversity in relationships. The chapter also shows how feminist and queer theorists have presented contested views on love, and analysed love as both liberating and oppressive. It also focuses on how love can be an important dimension of the same-sex marriage debate.

Intimacy, social change and globalisation

Jamieson (2011) argues that the cultural celebration and use of the terms 'intimacy' is not universal, but she argues that the practices of intimacy are present in all cultures. Jamieson describes intimacy as characterised by closeness, she maintains:

> The quality of 'closeness' that is indicated by intimacy can be emotional and cognitive, with subjective experiences including a feeling of mutual love, being 'of like mind' and special to each other. Closeness may also be

> physical, bodily intimacy, although an intimate relationship need not be sexual and both bodily and sexual contact can occur without intimacy.
>
> *(Jamieson, 2011: 1.5)*

Sociologists have been consumed with the relationship between intimacy and the nature of the self as shown in the work of Giddens (1992), Bauman (1995), Beck and Beck-Gernsheim (1995) and Castells (1997). Jamieson identifies Giddens's (1992) work as being the most explicit in its links between intimacy and the self. In Giddens's view, 'intimacy is built through a dialogue of mutual self-disclosure between equals, revealing inner qualities and feelings, simultaneously generating a self-reinforcing narration of the self' (Jamieson, 2011: 1.5). However, Jamieson notes that the emphasis on individualisation as a driver of change has been challenged (Brannen and Nielson, 2005; Bjornberg and Kollind, 2005; Charles et al., 2008; Crow, 2002; Duncan and Smith, 2006; Irwin, 2005; Jamieson, 1998, 1999; Smart and Shipman, 2004; Smart, 2007).

Non-Western countries have often followed a pathway that is a compromise between individualism and traditional patterns of responsibility. Jamieson (2011: 1.6) argues that:

> Research on Asian cultures experiencing significant economic development and growth in consumerism often found young people steering a middleway, both adopting practices associated with individualism – couples setting up their own homes, rather than living in three-generational families, choosing marriage partners rather than arranged marriages – and continuing to take collective responsibility for the elders and ancestors (Croll, 2006; Hansen and Pang, 2010; Ikels, 2004; Janelli and Yim, 2004; Quah, 2008; Yan, 2000).

Jamieson addresses practices of intimacy globally, and she maintains that anthropologists have reported on love and intimacy on every continent (see Cole and Thomas, 2009; Jankowiak, 2008; Hirsch and Wardlow, 2006; Li, 2008; Mody, 2008), and she maintains that whereas romantic passionate love relationships between couples have been asserted, the universality of friendship as a form of intimacy is a more contested domain.

She cites the work of Roseneil and Budgeon (2004) (see below) and Weeks (2007) as highlighting the issue of friendship relationships as replacing couple relationships in gay and lesbian relationships. She also cites examples from Europe, China, East Africa and Brazil to show that friendships involving affect and intimacy, or at least empathy, can be found globally.

Jamieson makes the important point that love is often conceptualised as an emotion (see Evans, 2003; see also Brooks, 2017), an embodied affect, and as she states, love in this usage becomes an attribute of a person rather than a quality of connection between people.

Theorists have also been interested in the relationship between the self and intimacy. The work of Michel Foucault (1978) has been significant, and while familial relationships are acknowledged in his account of how selves are shaped, it is the role of power, knowledge and discourse in his analysis of Western cultures that is seen as significant (see Brooks, 2017). Other social theorists such as Illouz (2007) focus more narrowly on the relationship between intimacy and the media (see Brooks, 2017):

> Some commentators focus more narrowly on the dominant messages of the mass media. Illouz, for example, like Bauman suggested that an emphasis on intimacy in the context of a voracious consumer culture leaves relationships: 'dispassionate, rationalised, and susceptible to crass utilitarianism' (Illouz, 2007: 109, see also Bauman, 2003).
>
> *(Jamieson, 2011: 4.4)*

Jamieson also raises the issue of the intersection of gender equity within relationships of intimacy, and she points out that practices of intimacy are not in themselves 'automatically democratising or dismantling of patriarchal arrangements'.

In sociology, the emphasis on diversity and the breakdown of heteronormativity has come from a range of theories, including postmodernism, black and minority ethnic feminism and lesbian and gay theorists. An example of this is found in the work of Roseneil and Budgeon (2004: 136). They develop the issue of intimacy and the way it has been traditionally understood in terms of diversity within sociology. They argue that these developments 'leave unchanged the heteronormativity of the sociological imaginary; and second they are grounded in an inadequate analysis of contemporary social change'.

They argue that sociological and anthropological research shows that 'friendship as both a practice and an ethic, is particularly important in the lives of lesbian and gay men' (see Preston and Lowenthal, 1996; Roseneil, 2000; Weeks et al., 2001). Roseneil and Budgeon (2004: 137–138) maintain that: 'Networks of friends which can include ex-lovers form the context within which lesbians and gay men lead their personal lives, offering emotional continuity, companionship, pleasure and practical assistance'.

Lesbians and gay men construct a set of relationships outside the mainstream heterosexual family. Roseneil (2000) and Weeks et al. (2001) maintain that there is a blurring of boundaries and movement between friendships and sexual relationships which characterises contemporary lesbian and gay intimacies. As Roseneil and Budgeon (2004: 138) state: 'Friends become lovers, lovers become friends, and many have multiple sexual partners of varying degrees of commitment (and none). Moreover, an individual's "significant other" may not be someone with whom she or he has a sexual relationship'.

They also argue that sociological research shows more broadly that definitions of intimacy has been undergoing significant change within the context of broader social change. Castells (1997), Giddens (1992) and Beck and Beck-Gernsheim (1995, 2002) all have discussed changing meanings and practices of love and relationships and changes in these. Roseneil (2000) goes much further and argues that:

> we are currently witnessing a significant destabilization of the homosexual/ heterosexual binary which has characterized the modern sexual order. She suggests there are a number 'queer tendencies' at work in the contemporary world, which are contributing to the fracturing of the binary [heterosexual/ homosexual]. More significantly the decentering of heterorelations, both socially and at the level of the individual.
>
> *(Roseneil and Budgeon, 2004: 140)*

She says there is a trend towards the normalisation of the homosexual in most Western nations, as there are progressive moves towards the equalisation of legal and social conditions for lesbians and gay men.

Theorising 'non-standard intimacies'

Two theorists who have developed the idea of 'non-standard intimacies' are Lauren Berlant and Michael Warner (2000). They make the point that public culture defines love and intimacy as closely tied to familialism; so does sociology, which tends to see the family as central in relationships. They argue that sociology continues to marginalise the study of love, intimacy and care beyond the family. Their research into the heteronormative public culture in the US 'constructs belonging to society through the "love plot of intimacy and familialism" restricting "a historical relation to futility . . . to generational narrative and reproduction"' (Berlant and Warner, 2000: 318).

Berlant and Warner argue that 'non-standard intimacies . . . created by those living non-normative sexualities pose a particular challenge by those living non-normative sexualities pose a particular challenge to a discipline which has studied intimacy and care primarily through the study of families'(Roseneil and Budgeon, 2004: 137).

The heterosexual couple, and particularly the married, co-resident heterosexual couple with children, is no longer the norm in Western societies. This is the result of a number of factors: the dramatic rise in divorce rates; the increase in the number of births outside marriage; the increase in 'single by choice' mothers; the increase in the number of single-parent households; and the increase in the number of women not having children.

Intimacy and emotion work in lesbian, gay and heterosexual relationships

The work of Umberson et al. (2015: 542) builds on this acknowledgement of heteronormativity by some sociologists and social and cultural theorists, and merges a 'gender-as-relational perspective – that gender is co-constructed and enacted within relationships – with theoretical perspectives on emotion work and intimacy to frame an analysis of in-depth interviews'. They found that emotion work directed towards minimising and maintaining boundaries between partners is key to understanding intimacy in long-term relationships.

They argue that emotion work is a common strategy for enhancing intimacy between partners, and, in heterosexual relationships, women are much more likely than men to do this kind of emotion work (Elliott and Umberson, 2008). Umberson et al. (2015), in their research, examine the possibility that intimacy is enacted and experienced by men and women in different ways depending on whether they are in a relationship with a man or a woman.

They found that compared with men, women express a desire for more emotional intimacy, and report more frustration with levels of intimacy in their relationships (Peplau, 2001). Women also work harder to promote emotional intimacy in emotional relationships by urging communication and the sharing of personal feelings.

Umberson et al. (2015: 543) comment that:

> Women in heterosexual relationships are more likely than men to repress their own feelings (a form of emotion work) to foster intimacy and their partner's well-being (Elliott and Umberson, 2008; Erickson, 2005; Thormeer et al., 2013). The experience of intimacy in heterosexual relationships is further characterised by a gendered view of the link between emotional intimacy and sexual interactions.

In addition, research suggests that women are more likely than men to view emotional intimacy as essential for sexual interactions. Elliott and Umberson (2008) show that in heterosexual relationships, gender differences in emotional expression, sexual expression and desire often contribute to relationship strain and conflict, sometimes compromising intimacy.

Research on lesbian couples has highlighted the lack of boundaries between lesbian partners. Compared with men in same-sex relationships, women in same-sex relationships (similar to women in heterosexual relationships) put greater emphasis on the relationship between emotional intimacy and positive sexual interaction.

Men in same-sex relationships are more likely to approve of and have sexual relationships outside their committed relationships and to separate sex and emotional intimacy (Peplau and Fingerhut, 2007).

Umberson et al. (2015: 543) maintain that 'Non-monogamous gay couples sometimes establish sexual contacts that set rules against emotional intimacy with sexual partners outside the relationship (Peplau and Fingerhut, 2007)'. They argue that while studies have compared the relationship experiences of lesbian, gay and heterosexual couples (e.g. Julien et al., 2003; Kurdeck, 2006), they have not focused explicitly on meanings and experiences of intimacy;

neither have they considered how partners work to promote intimacy in their relationships.

It is well understood that cultural ideas about women and men in relationships (Elliott and Umberson, 2008; West and Zimmerman, 2009) are likely to be reflected in their experiences of intimacy. It is argued that the 'gender as relational' approach suggested by Umberson et al. (2015: 544) has the 'power to advance our understanding of gendered experiences in relationships by emphasizing that intimacy might be experienced and expressed differently by men and women depending on their social contexts and with whom they are interacting' (Goldberg, 2013). As Elliott and Umberson (Umberson et al. 2015: 544) note:

> The gender-as-relational approach allows us to consider how same-sex couples might queer intimacy (Goldberg, 2013; Oswald et al., 2005). *Queering intimacy* means challenging heteronormative gendered view of intimacy (e.g. women want more intimacy, men resist intimacy, and partners have different beliefs about the meanings and experiences of intimacy). Same-sex couples may diverge from heteronormative patterns of intimacy and inequality. They may do so in part by engaging in different types of work to promote intimacy and influence boundaries in their relationships. Alternatively same-sex couples may enact intimacy in ways that parallel heteronormative scripts of different-sex partners.

Much has been written about Hochschild's (1979) concept of 'emotion work' (see Brooks, 2006, 2010; Brooks and Devasayaham, 2013; Brooks and Simpson, 2012) and Hochschild originally used the term 'emotion work' to refer to efforts involved in managing personal emotions in an attempt to promote positive emotions in others. She suggested that 'emotion work' would be most prevalent in the context of intimate relationships, and that emotion work would be strongly gendered as a result of gendered expectations.

Umberson et al. (2015) show that studies of heterosexual couples have documented that women undertake substantially more emotion work than men in an attempt to bolster self-esteem in their partners as well as positive emotions. Elliott and Umberson (2008: 403) 'found that emotion work extends to sexual interactions, whereby individuals alter their own sexual desires to conform to those of their own partner "in an effort to reduce marital conflict, enhance intimacy, [and] help a spouse to feel better about himself or herself"'.

Based on research undertaken by Umberson et al. (2015) they found that compared with men, women devoted much more discussion to the importance of minimising boundaries between partners in an effort to promote intimacy. Approximately half of the women involved in lesbian and heterosexual relationships emphasised the importance of minimising boundaries between partners to sustain intimacy compared, with approximately one-fifth of the men in gay and heterosexual relationships.

Further, the findings of the research conducted by Umberson et al. (2015: 545) present a nuanced and sophisticated understanding of relationships, as follows:

> Scholars have called for more attention to a queer perspective in the study of relationships and families – that is, to move beyond a heteronormative focus based on a gender binary (Goldberg, 2013; Oswald et al., 2005) Our finds suggest a blend of gender conformity and contestation in same sex relationships. For example lesbian couples adhered to traditional feminine (gendered) systems of intensive emotion work and a desire for emotional intimacy, yet they contested heteronormative views of partner discordance in the desire for intimacy and specialization in emotion work directed towards intimacy. Gay couples adhered to traditional masculine (gendered) systems of boundaries (e.g. emotional autonomy and independence) in the context of their committed long-term relationships, yet they contested heteronormative expectations when they carefully monitored a partner's need for emotional support and then stepped in to provide that support. Our findings indicate that the gendered relational context of lesbian and gay couples create unique intimacy systems that sustain their relationships overtime. Furthermore these systems queer our understanding of intimate relationships by diverging from those of heterosexual couples. In this sense, same-sex couples occupy 'creative spaces . . . where new constructions get crafted and old ones are remade' Oswald et al., 2005: 148). Overall our analysis of lesbian, gay and heterosexual couples in long-term relationships suggests multiple successful pathways to intimacy and relationship longevity.

Feminist and queer perspectives on love and intimacy

Much has been written within feminist narratives on love and intimacy. In this overview of the diverse views of feminists globally, I draw on the work of some of the Australian feminist and queer theorists who have illuminated the debates within an Australian context.

Possibly one of the most 'shrill' Australian feminist theorists has been Germaine Greer, and her initial contribution to feminist debate on love and intimacy, captured in *The Female Eunuch* (1970), was ground-breaking. As with much feminist discourse of that era, Greer 'excoriated' love in the following way:

> Love, love, love – all the wretched cant of it, masking egoism, lust, masochism, fantasy under a mythology of sentimental postures, a welter of self-induced miseries and joys, blinding and masking the essential personalities in the frozen gestures of courtship, in the kissing and the dating and the desire, the compliments and the quarrels which vivify its barrenness.
>
> *(The Female Eunuch, 1970) (cited in Teo, 2014a)*

Greer's (1970) view was one that was internationally supported by other leading second-wave feminists, including Simone de Beauvoir (1972) and Shulamith
Firestone (1972). De Beauvoir (1972: 653) maintained that 'love enslaved women',
while Firestone (1972: 121) claimed that love 'is the pivot of women's oppression today'. This view of love is affirmed in Mary Evans's (2003) book *Love: An
Unromantic Discussion.*

A range of feminist theorists were also critics of marriage as oppressive and dis-
advantageous to women (Lake, 2013; Summers, 1975). Teo (2017a: 19) also shows
how feminist scholarship on marriage in Australia, which includes writers in the
area of postcolonial and Indigenous studies, emphasise the 'deleterious effects of
these relationships on Aboriginal individuals and communities' (see Grimshaw,
1994, 2002; McGrath, 2002; Ellinghaus, 2002, 2003; Conor, 2013).

Jonasdottir and Ferguson (2014) look at the different ways in which feminist
and other theorists have approached love. In their book *Love: A Question for
Feminism in the Twenty-First Century*, Jonasdottir and Ferguson argue that their
focus is on the material practices and embodied experiences of love, power and
domination in order to move towards liberation. They thus see love as a mecha-
nism towards liberation. They point out that traditional feminist theories of love
have concentrated on heterosexual romantic love as a reflection of patriarchal
ideology and male dominance. In fact, some of the classic feminist writers were
radical opponents of women in conventional relationships of dependence (Wollstonecraft, 2004; Flexner, 1972).

They also point out that Anglo-American feminist theorists returned to the
subject of sexual love in the 1990s and focused on 'romantic love', 'romance' and
'romantic utopia' (Pearce and Stacey, 1995; Illouz, 1997). As shown elsewhere
in this book, there was also a focus on psychoanalytic theories of sexual desire
around the same time, particularly linked to film and discourse analysis, as outlined by Jonasdottir (2014: 16):

> The main paradigmatic horizons and theoretical frames, found useful for
> feminist analyses from the 1960s onwards (such as historical material-
> ism, different strands of psychoanalytic theory, French linguistic struc-
> turalism. followed by poststructuralist theories), seemed not to have any
> justifiable place for love as such other than as an epiphenomenon. . . .
> Theoretically, therefore love was seen as a form of labor (or work), or as
> an ideological delusion; an idealized image of (inhibited) sexual desire;
> or a symbolic or discursive element performing disciplinary power. In
> each case the explanatory, interpretative or liberating power of love was
> seen as very limited.

As Ferguson and Jonasdottir (2014: 16) note, 'An emphasis on power and con-
textual differences between women by race, class, ethnicity and sexuality also
led poststructuralists and postmodernist theorists to reject any identity politics
based on a supposed essentialist epistemological standpoint of womanhood' (see
Brooks, 1997 for a complete summary).

Other social theorists link critiques of love with theories of modernity, as shown in Chapter 1, particularly theories of individualisation and democratisation of heterosexual love (Beck and Beck-Gernsheim, 1995; Giddens, 1992). Others focus on the theme of 'emotional capitalism' through linking love and consumerism (Evans, 2003; Kipnis, 2003; Illouz, 2007, 2012).

Traditional feminist perspectives on love are captured in the work of Kipnis (2003) and Evans (2003). Kipnis, in *Against Love: A Polemic*, argues against love as being the most efficient kind of social control, while Evans in *Love: An Unromantic Discussion* argues that love is counterproductive for women, particularly in its 'romanticized and commercialized forms' (Evans, 2003: 143). Jonasdottir (2014: 14), by contrast, is one of a number of feminists who see love as a 'creative power, a productive force with (at least a potential) positive value, conceptualized and theorized beyond the constraining power of (an assumed) delusion or ideology called "romantic love"'.

An interesting distinction can be made between the modernist social theorists of love such as Luhmann (1986), Beck and Beck-Gernsheim (1995) and Giddens (1992) who concentrate 'on the specific realm of intimacy, on love as a historically changing code that regulates that "quite normal chaos" in intimate relations, particularly sexual relations' (Jonasdottir, 2014: 19). This is distinct from Illouz's work which focuses on the 'changing "structure of the romantic self", and on romantic love as a culturally constructed commodity a product she claims has become an "intimate indispensable part" of the capitalist market' (Jonasdottir, 2014: 19).

Illouz's work is more interesting and insightful than most feminist theoretical contributions, and argues that patriarchy is the pivotal element that structures inequality between the sexes and the privileging of heterosexuality. While of course this is well established, Illouz goes much further and argues more imaginatively in her analysis, in showing that 'it alone cannot explain the extraordinary grip of the love ideal on modern men and women' (Illouz, 2012: 5–6, 1997: 2).

Eva Illouz (see Chapter 9 and Brooks, 2017) has provided a sustained critique of love, sex and marriage in *Why Love Hurts: A Sociological Explanation* (2012):

> Illouz remains committed to love as a central idea of modernity, she champions its egalitarian optimism and its ability to subvert patriarchy, however she acknowledged that love is also a source of much misery. This misery, she argues, stems from the 'institutional arrangement' surrounding it. Love is played out in 'the market place of unequal competing actors' in which some people, mostly men, are able to 'command a greater capacity to define the terms in which they are loved by others'.
>
> *(Grossi, 2013)*

As I have shown elsewhere (Brooks, 2017), Illouz highlights how men command the sphere of commitment. Illouz maintains that: in relation to commitment, men are less likely to desire marriage and family, because these are no longer sites of control and domination. Men now measure success not according to a successful commitment but rather success on the sexual market. As such, men wish to remain uncommitted for as long as possible (Grossi, 2013).

Another theorist who has a fairly pragmatic view of love is Marilyn Friedman (2003) in her book *Autonomy, Gender, Politics*:

> Friedman argues that the features of merger experienced within romantic love are that the needs and interests of each person become entwined and pooled together, couples feel each others highs and lows; there is a mutual consideration and awareness. . . . Friedman does not necessarily consider these features as always already negative, but they can represent a significant reduction in personal autonomy, and this is more dangerous for women than for men.
>
> *(Grossi, 2013)*

Another feminist commentator, Wendy Langford (1999), in *Revolutions of the Heart: Gender, Power and the Delusions of Love*, also disputes the idea that the ideology of love can be positive. As Grossi maintains:

> She argues that while the idea that love has spread principles of justice and fairness is an attractive and optimistic view, it is empirically unsustainable and conceptually misguided. She claims 'Love does not merely fail to give us what we desire, but in doing so, compounds painful feelings of dissatisfaction and low self-esteem'.
>
> *(Grossi, 2013)*

Thus, she sees the results as negative. Other feminists go further and see love as a site of resistance, transformation and agency. Illouz sees love as both egalitarian and subversive (see Chapter 9 and see Brooks, 2017). Similarly Pearce and Stacey (1995) in *Romance Revisited* argue: 'that love retains its ability to liberate women from patriarchy because of its narrativity'. Grossi (2013) shows that 'They argue that an engagement with the narrative of romance enables women to facilitate the "rescripting of other areas of life"'. In setting out these perspectives it is clear that love can be seen to be both liberating and oppressive.

Love in same-sex relationships and marriage

Another important dimension of these debates is the work of queer theorists. Johnson (2005) in his book *Love, Heterosexuality and Society* argues that 'love is constructed around scripts of nature and linked scripts of heterosexuality, marriage, procreation and family. This renders invisible love that exists in relationships that do not adhere to these scripts' (Grossi, 2013). In other words, alternative scripts – such as homosexuality and same-sex marriage – remain invisible.

Sean Salvin's (2009) in '"Instinctively, I'm Not Just a Sexual Beast": The Complexity of Intimacy among Australian Gay Men', argues that: 'gay men have to fight for the recognition of the relationships they have. Open relationships, casual sex with regular partners, multiple sex partners, can all, according to Salvin represent love' (Grossi, 2013). The pattern of gay relationships must contest the normative structures of heterosexual couples, argue Bell and Binnie (2000) in their book *The Sexual Citizen: Queer Politics and Beyond*. They maintain 'that love must move away from the couple and include non-monogamy, polyamory and episodic sexuality' (Grossi, 2013). In other words, the normative structures around heterosexuality render invisible the nature of same-sex relationships. 'Johnson and other queer theorists, then argue for dismantling a number of restrictive associations that define love: the breakdown of the binary of love and sex; the breakdown of the connection between love and marriage, family and procreation' (Grossi, 2013).

One of the most significant feminist theorists who addresses queer politics is Lauren Berlant (see Brooks, 2017). In *Love, A Queer Feeling*, Berlant (2001) argues that love is ultimately a site of optimism or change for a transformational environment. Berlant maintains that love needs to be connected more directly to agency, and as something that exists within different structures, both within and outside normative structures of heterosexuality.

> Berlant argues that when queer thought enters the discourse of love, it must not teach 'that we are all alike and compelled to repeat our alikeness intelligibly, but by teaching some of what we learned about love, under the surface, across the lines, around the scenes informally'.
>
> *(Berlant, cited in Grossi, 2013)*

Berlant maintains that 'queering love' is only fully achieved when it operates outside established institutions. She maintains that love can only avoid oppression when it rejects established normative patterns and rejects rules and barriers that limit intimacy within established structures.

Grossi (2012) argues that Berlant's work has important implications for the role of love in the same-sex marriage debate. In an article entitled 'The Meaning of Love in the Debate for Legal Recognition of Same-Sex Marriage in Australia', Grossi shows how love as a concept provides an important dimension in the debate around same-sex marriage.

Drawing on Berlant (2001), the issue of love within same-sex marriage can and should be overcome, because love is ultimately a site of optimism, change and transformation. As Berlant (2001: 448) notes, 'love approximates a space to which people can return, becoming as different as they can from themselves without being traumatically shattered; it is a scene of optimism for change, for a transformational environment.'

However, as already noted, to achieve this, love must be connected to agency and as existing outside 'heterosexual scripts':

> 'Queering' love for Berlant is achieved when it lives up to its promises of existing outside of established institutions, when it challenges all rules connected with it which presume to establish principles for living. In

other words when love delivers what it promises: intimate relationships that are free of oppressive and traditional forms. Berlant is interested in love because of its potential for freedom and possibilities and rejects it as a means of establishing rules and barriers.

(Grossi, 2012: 499)

Conclusion

Social relationships and the nature of love and intimacy are significantly diversified, and concepts of marriage and the family have overcome traditional conceptualisation around heteronormativity. Some social theorists, such as Jamieson, have focused on social change and globalisation, challenging models of intimacy which have relied on individualisation as developed in the work of Giddens. Jamieson looks at non-Western perspectives on relationships and the family as providing different patterns. Roseneil and Budgeon and Weeks highlight how friendship relationships are significant, in contrast to romantic passionate love relationships. As Roseneil and Budgeon show, boundaries between friendships and sexual relationships provide greater fluidity in lesbian and gay relationships. Berlant and Warner's interesting work has been defined as 'non-standard intimacies'. The chapter also focuses on 'emotion work' and emotional intimacy in lesbian, gay as well as heterosexual relationships in a nuanced and sophisticated analysis of gender as relational. The chapter also shows how feminist and queer theorists have presented contested views on love and analyse love as both liberating and oppressive. It also focuses on how love can be an important dimension of the same-sex marriage debate.

3

BEING YOUNG AND IN LOVE

Introduction

This chapter looks at how young women and young people more generally exercise choice in their lives and how this impacts on their intimate relationships. The chapter draws on a wide range of contemporary studies, including research undertaken by the Pew Research Center (2014) in the United States, who investigate the relationship between *Teens, Technology and Romantic Relationships*. The Report examines the role that social and digital media play in romantic relationships of youth in America. A wide range of sources show young people are negotiating romance and intimacy in the context of wider pressures in their lives. Helen Willmot (2007) explores the relationship between young women, opportunities through education and employment and how they construct love and intimacy. Richtel (2012) looks at how new technology plays into relationships with younger people and their response to it. Sharon Elley (2015), in a fascinating study of 'laddishness' in higher education, looks at the growth of inappropriate, abusive and misogynistic behaviour in universities. Elley considers the implications of this for gender politics in higher education and the growth of masculine behaviour. Finally, the chapter considers the model of youthful relationships put forward by Kathleen Bogle. She maintains that young people in higher education highlight a pattern of 'hooking up' in their association with their peers which differs significantly from earlier patterns of dating. The chapter also looks at the historical impact of romance comic books on love and romance among young people in Australia, and how many of the stories and images were adapted from an American context to suit an Australian setting.

Teens, technology and romantic relationships

A report was conducted by the Pew Research Center (2015) between 2014 and 2015 which examines American teens' digital romantic practices. The age range was between 13–17. The report focuses on 'Understanding the role social and digital media play in romantic relationships is critical, given how deeply enmeshed these technology tools are in the lives of American youth (Pew Research Center, 2015)'. The research covers the following areas: American teens' digital romantic practices; social media, gender and flirting; girls are more likely to be targets of these practices; boys' view of social media; teen dates and public demonstrations of affection; texting and teens; and harmful digital behaviour and teen daters.

The Pew Research Center (2015) report comments that 'From flirting to breaking up, social media and mobile phones are woven into teens' romantic lives'. The study reveals that:

> . the digital realm is one part of a broader universe in which teens meet, date and break up with romantic partners. Online spaces are used infrequently for meeting romantic partners, but play a major role in how teens flirt, woo, and communicate with potential and current flames.

The main findings of the report are as follows:

- Relatively few American teens meet a romantic partner online: a quarter, 24 per cent of teen daters or roughly 8 per cent of all teens, have dated or hooked up with someone they first met online;
- Teens are far less likely to have embarked on a romantic relationship that started online. A majority of teens with dating experience, 76 per cent, say they have only dated people they met via offline methods;
- Of those who met a partner online, the majority met on social media sites, and the bulk of these met on Facebook;
- Social media is the top venue for flirting. While most romantic relationships do not start online, technology is a major vehicle for flirting and expressing interest in a potential partner;
- In relation to flirting, the report shows the following:

  - 55 per cent of all teens 13–17 have flirted or talked to someone in person to let them know they are interested;
  - 50 per cent of teens have let someone know they are interested in them romantically by befriending them on Facebook;
  - 47 per cent have expressed their attraction by liking on social media;
  - 32 per cent have sent flirtatious messages;
  - 11 per cent have made them a music playlist;
  - 10 per cent have sent them flirty or sexy pictures or videos of themselves;
  - 7 per cent have made a video of themselves for them.

More advanced and sometimes overtly sexual suggestive online behaviours are often exhibited by teens who have prior experience in romantic relationships: 63 per cent of teens with dating experience have sent flirtatious messages to someone they were interested in; just 14 per cent of teens without dating experience have done so; 23 per cent of teens with dating experience have sent sexy or flirty pictures to someone they were interested in, compared with just 2 per cent of teens without dating experience.

The report also found that girls are more likely to be targets of uncomfortable flirting tactics. Not all flirting behaviour is appreciated or appropriate. One quarter, or 25 per cent, of all teens have unfriended or blocked someone on social media because that person was flirting in a way that made them uncomfortable. Just as adult women are often subject to more frequent and intense harassment online, teen girls are substantially more likely than boys to experience uncomfortable flirting with social media environments.

The report states that boys are a bit more likely than girls to view social media as a space for emotional and logistical connection with their significant other: 65 per cent of boys say social media makes them feel more connected with what is happening in their significant other's life, compared with 52 per cent of girls. Some 16 per cent of these boys report that these platforms make them feel more connected. In addition, 50 per cent of boys say social media makes them feel more emotionally connected with their significant other, compared with 37 per cent of girls. Some 13 per cent of boys felt a lot more emotionally close.

Interestingly, teen daters like being able to publicly demonstrate their affection and show support for others' romantic relationships. As the Pew Research Center (2015) report indicates:

> For some teens, social media is a space where they can display their relationships to others by publicly expressing their affection. . . . More than a third, 37 percent of teens with relationship experience have used social media to let their partners know how much they like them. They say social media is a place where they can show they care.

In addition, teens also use social media to express public support of other people's romantic relationships. Nearly two-thirds, or 63 per cent, of teens with dating experience have posted or liked something on social media as a way to indicate support of their friends' relationships. Girls are especially likely to support friends' relationships on social media: 71 per cent of girls with dating experience have done so compared with 57 per cent of boys.

Other aspects of the report which are of note include:

- Many teens in romantic relationships, 85 per cent, expect their significant other to be in daily communication, at least once a day or indeed more often; 15 per cent expect to check-in hourly; 38 per cent expect to check in every few hours; 35 per cent expect to check in once a day;
- Texting, voice calls and in-person hanging out are the main ways teens spend time with their significant others;
- Teens consider the text message breakup to be socially undesirable, but a sizeable number of teens with relationship experience have been broken up with or have broken up with others using text messaging.

A relatively small number of teen daters engage in potentially controlling or harmful digital behaviour to a partner or ex-partner. Such behaviour includes the following:

- 11 per cent of teen dates have accessed a mobile or online account of current or former partner;
- 10 per cent of teen daters have modified or deleted their partner's social media profile;
- 10 per cent have impersonated a boyfriend or girlfriend or ex in a message;
- 8 per cent have sent embarrassing pictures of a current or former partner to someone else;
- 4 per cent of teens have downloaded a GPS or tracking program to a partner's device without their knowledge.

A small share of teen daters have experienced potentially abusive or controlling behaviour by a current or former partner:

- 16 per cent of teen daters have been required by a current or former partner to remove former girlfriends or boyfriends from their list on Facebook, Twitter or other social media;
- 13 per cent of teens with dating experience report that their current or former partner demanded that they share their passwords to email and internet accounts to them;
- 11 per cent of teens with relationship experience report that a current or former partner had contacted them on the internet or on their cell phone to threaten to hurt them;
- 8 per cent of teen daters report that a current or ex-partner used information posted on the internet against them to harass or embarrass them;
- 22 per cent of teens with relationship experience have had a partner use the internet or a cell phone to call them names, put them down or say mean things to them;
- 15 per cent of teen daters report that a current or former partner spread rumours about them using a digital platform like mobile phones or internet.

(drawn from the Pew Research Center, 2015)

Intimacy, diversity and young women as agents of change

There are clearly gender differences in the responses of young people to intimacy and relationships, and within different gender groups there are also differences based on class, ethnicity and educational opportunity. Willmot (2007) in an article entitled: 'Young Women, Routes through Education and Employment and Discursive Constructions of Love and Intimacy' explores the interaction between young women's educational and subsequent occupational experiences and their intimate relationships with men, drawing on 30 in-depth interviews with young women. She looks at different routes through education and employment and shows how this positions them in different ways in relation to intimate relationships. She shows how this opens up two discourses, one conducive to romance and one to an opposing discourse of contingency, resulting in different practices. Willmot (2007: 447) shows that:

> Within the sphere of intimacy 'One of the key issues of the 1990s has been precisely the attempt to move from recognition to normalization of diversity' (Weeks, 2000: 173). By the 1990s, there was recognition of 'alternative families', differentiated by class, ethnicity, race, single parenthood and chosen lifestyles, including cohabitation (Weeks, 2000: 212).

It was also clear that changes in women's lifestyles and choices were having a fundamental impact on their relationships:

> Young women are held by theorists such as Giddens and Beck and Beck-Gernsheim to be at the forefront of the shift towards such choice and subsequent diversity within intimate relationships, due to the increased educational and subsequent occupational opportunities available to them, which free them from traditional gender roles within intimate relationships with men.
>
> *(Wilmot, 2007: 446–447)*

Arnett (2004) states that before 1970, the typical 21-year-old would be married, caring for a child and making serious decisions about their life at an early age. However, today this is very different, as many young people have individualised priorities predominantly focusing on advancing their education and career prospects, rather than choosing marriage and commitment at a young age.

Wilmot's contention supports the data published by the Pew Research Center (Livingston, 2015) which claims 'among women ages 40–50, the median age at which those with a Masters degree or more, first became mothers now stands at 30. In comparison, the median age at first birth for women with a high school diploma or less is just 24'.

Wilmot's research shows that there is a connection between a discourse of contingency and certain experiences. She undertook an empirical study based on interviews with 30 young women. She argues that the different routes through education and employment position the women in different settings which are either conducive to a discourse of romance and one of an opposing contingency. Thus, the women are exposed to different experiences. All the women defined themselves as heterosexual.

These experiences are the pursuit of a higher education and subsequent jobs which tend to involve geographical relocation(s) and living in shared/student houses. As Wilmot (2007: 459) comments:

> Within the sample . . . then there are two groups of young women who have discursively constructed intimate relationships very differently

and whose intimate relationships have been conducted in very different settings, linked largely to their different routes through education and employment. For those who left full-time education by the age of 18, intimate relationships tended to be conducted in a setting characterized by geographical mobility, or at least the potential for this, and shared/ student houses.

Thus, as Wilmot shows, there are differences in the discourses of love and intimacy facilitated by different routes through education and employment. Leaving full-time education by the age of 18 tended to mean that relationships were conducted within a situation of geographical stability and living with parents. Higher education meant that intimate relationships were conducted in a situation of geographical mobility, which encouraged contingency. This is similar to Gidden's (1992: 61) conceptualisation of 'confluent love', defined as clashing with romantic notions. As Wilmot shows Giddens argues that that romantic ideals tend to fragment under young women's increased autonomy. He links confluent love to young women's greater autonomy derived from equal opportunities in education. In the study contingency was associated with the pursuit of higher education, thus the findings support such a conclusion.

Elley (2015), in 'Confronting a Culture of "Laddishness" and Riskiness in Higher Education', examines male behaviour in higher education, maintains that UK universities are witnessing considerable controversy surrounding male students' behaviour in what has been termed an increased 'culture of laddism', sexism, harassment and violence. Elley reports that in 2013, one campus became infamous due to the closure of a 'Tequila' club event which hosted a 'Violate a Freshers' Night. Elley (2015: 2) provides a contextualisation of the current situation:

> A generation surrounded by arguably . . . instant information, immediate gratification and quick fixes, many students are ill-prepared for university and academic life, in its current state, with both young men and women students vulnerable to the increased excessiveness, intensification and acceleration found in higher education and general life. Expected (and desiring) to 'fit in', 'stand out', 'take risks' and make 'sensible choices', it is unsurprising then that students may increasingly seek the means to momentarily escape (i.e. through substance use) and push the boundaries of acceptability via laddism, pleasure and risk-seeking behaviours which are somewhat institutionally escalated.

Elley points out that often laddism, sexism and abuse targeted at women and between lads is about a hierarchical ordering of masculinities attached to class privilege, taste and distinction as students reinstate boundaries. Outside of the university context, she states, this discrimination would not be tolerated.

Traditionally, as Phipps and Young (2014) theorise, higher education was associated with hierarchy, privilege and 'male' competition. However, young women are increasingly open to opportunities and marketisation of neoliberal universities. *The Guardian* defines lad culture as follows: 'laddism is an equal-opportunity oppressor – racism, classism, homophobia and transphobia are all part of its portfolio, but the viciousness of its sexism, reflects a conviction that women need to be put in their place' (Phipps and Young, 2014).

The National Union of Students (NUS) (2010) has called for campus-wide action, with Student Unions banning inappropriate laddish, abusive behaviours and the playing of chart hit songs like 'Blurred Lines' (2013) for its misogynistic lyrics. Current research reveals female student accounts of the alarming negative impact that 'laddism' has on their educational experiences and well-being (NUS, 2010; Phipps and Young, 2014, 2015). Laddish behaviour overlaps with laddism, which loosely means a group or 'pack mentality' shared through sporting societies, heavy drinking and sexist and homophobic banter. It is often connected to the sexualised objectification of women and pro-rape attitudes, including sexual harassment and violence.

The NUS (2010) *Hidden Marks* report examined the experience of female students, and found one in seven female students had experienced some form of sexual or physical assault, with 68 per cent of students experiencing verbal or non-verbal harassment in or around their universities. The study found over a third of women felt unsafe visiting universities in the evening, and 60 per cent of students heard rape jokes on campus.

More widely, these issues link to casual 'hooking-up', partying, substance use (i.e. students waking up unable to recall whether they have consented to sex or been raped). As Elley (2015) points out, we are perhaps witnessing not a reaction to neoliberal principles and increased gender equality (see Phipps and Young, 2014), but the privileging of particular masculine behaviours and spaces where lads (re)claim lost power between lads as much as against women.

Being young and in love and sharing everything

Richtel (2012) maintains that the digital era gives rise to more intimate customs of sharing as young people express their affection for one another by sharing their passwords to email, Facebook and other accounts. As Richtel shows:

> The digital era has given rise to a more intimate custom. It has become fashionable for young people to express their affection for each other by sharing their passwords to e-mail, Facebook and other accounts. Boyfriends and girlfriends sometimes even create identical passwords, and let each other read their private e-mail and texts.

The stories of fall-outs of such sharing are extensive and damaging as Richtel shows:

> The stories of fallout include a spurned boyfriend in junior high who tries to humiliate his ex-girlfriend by spreading her email secrets; tensions

between significant others over scouring each other's private messages for clues of disloyalty or infidelity; or grabbing a cellphone from a former best friend, unlocking it with a password and sending threatening texts to someone else.

Young people are, of course, drawn to this kind of behaviour because parents and others might urge against it, so there really is little that parents and advisors can do to prevent such behaviour. The problems emerge when young people are subsequently bullied by others and pressures can mount into suicidal behaviour.

'Hooking up': sex, dating and relationships on campus

In an interesting study on relationships among college students in the US, Bogle (2008) discusses how the focus of relationships has moved from dating to 'hooking up'. Bogle (2008: 2) comments on the reasons for this:

> I want to suggest that two factors have been especially important in the demise of traditional dating on college campuses. First young people are postponing marriage. Age at first marriage is at an all-time high; the typical groom is 27. . . .; the typical bride is 25. Although today's men and women may be delaying marriage they are often sexually active from adolescence; the average intercourse is 17. Second a growing proportion of young people nationwide are spending the early years of their adult life on college campuses. From 1970 to 2000, enrolment in undergraduate institutions rose by 78 percent. Thus college has become an increasingly important setting for early sexual experiences.

Bogle shows that in 2001, a national study on college women's sexual attitudes showed that instead of dating, students were now 'hooking up'. She explains the context of this: the study defined a 'hook up' as 'when a girl and a guy get together for a physical encounter and don't necessarily expect anything further'. The response to this study was a media firestorm. Some of the concerns are as follows:

> Further, hooking up has been connected to an array of social problems such as binge drinking, drug abuse, and sexually transmitted diseases. In addition, feminist scholars have been concerned about the link between hooking up and sexual assault, while conservatives have linked hooking up to being raised by divorced parents.
>
> *(Bogle, 2008: 3)*

Bogle also refers to a second study conducted by the Institute for American Values which indicates that hooking up is a nationwide phenomenon that has largely replaced traditional dating on college campuses. Bogle (2008: 5) indicates:

This study examined the sexual attitudes and behaviours of college women across the country and found that hooking up was a common activity that dominates male-female relationships, believed hook ups occurred 'very often' or 'fairly often' on their campus, and 40 percent had personally engaged in a hook up encounter since coming to college. The researchers concluded that 'hooking up a distinctive sex-without commitment inter-action between college men and women is widespread on campuses and profoundly influences campus culture'.

Defining hooking up

Bogle acknowledges that it is difficult to define hooking up and that mean-ings differ and can include 'having sex', 'just kissing', 'making out' and 'fooling around'. The outcome of 'hooking up' is not really what the emphasis is about:

> After 'hooking up' someone can opt to ask for the other's phone num-ber or can try to make plans to meet somewhere in the future, but most students indicated that this is not the most common outcome. Instead students said that the most likely outcome of any particular hook up encounter is 'nothing', which means not hearing from the person again unless you coincidentally see him or her at another social event and decide to hook up again.
>
> *(Bogle, 2008: 28)*

As Bogle notes, 'hooking up' is an outgrowth of how college students socialise currently. Bogle (2008: 39) also suggests that 'hooking up' may have unintended consequences:

> less experienced college women may be sexual with someone with the hope that such behavior will lead to a relationship, may not suspect that their sexual availability decreases their chances of having the man pur-sue a relationship. One quantitative study confirmed what the upper class women I spoke with, believed that is, 49 percent of college students who engaged in sexual intercourse during a hook up encounter said they never saw the person again. Indeed, members of the campus culture had to *learn over time* the rules of the hook up script.

Young romance and comic books in Australia

The influence of the United States on patterns of romantic interaction and in the consumption of romance in both its products and its influences is shown throughout this book, and has been historically very significant in a number of countries. An interesting analysis of the influence of Americanisation on young people and romance in Australia is provided through the analysis of comic books.

Patrick (2017) shows that while the American-driven penetration of romance comics into Australia was condemned because of their risqué illustrations, which branded them as 'sex comics' promising 'torrid passion' to their readers, in fact the narratives were very 'safe'. As Patrick (2017: 224) notes:

> most romance comics espoused a conservative morality that guided read-ers through the complexities of romantic love and courtship. No matter how many tears were shed by their heroines, romance comics continu-ally emphasised that ultimate happiness was possible – but only through matrimony.

As shown elsewhere (see Teo, 2006, 2017b) the romantic comic book was part of a much wider influence of American popular culture on Australian society.

The target audience and markets for the romantic comic books is interesting and, as Patrick (2017: 225) shows:

> Simon and Kirby developed a prototype edition of their comic-book, *Young Romance* for Crestwood Publications in 1947. The debut issue firmly estab-lished the parameters of the romance comic-book genre, featuring fictional 'confession stories told by teenage girls' often 'plagued by guilt for such acts as falling in love with a delinquent' – which were happily resolved, leaving the girl 'thoroughly cleansed of sins' (Simon, 1990: 123).

In fact, Simon's target audience was an older audience, rather than the teenage high-school girl, but he was up against the wholesalers who liked the teen mar-ket. However, Simon maintained the focus on 'realistic adult stories'. Despite the female readership, the illustrations, which were stunning and often raunchy, were drawn by men.

Patrick shows that by 1949, there were 90 comic book titles published in Australia. Connell (1957) indicates that the circulation of comic books was 50 million by 1951 and 60 million by 1954. Connell *et al.*'s University of Sydney study *Growing Up in an Australian City: A Study of Adolescents in Sydney* provides an interesting socio-historical analysis of adolescents in Sydney in the 1950s.

> Provocative titles such as *Dramatic Love* . . . and *Intimate Confessions* invite the reader to share vicariously the affairs of young men and women experi-menting with their social relationships and expressing graphically their emotional reactions to the poignant and sometimes precarious circum-stances into which their experience may lead them.
>
> *(Connell et al., 1957: 156)*

Patrick makes the point that young women and girls were not the only audience for romance comics, and interest among boys grew as they became older. He also notes that there were reports indicating that romance comics were avidly read by older men as well. Patrick (2017: 234) makes the insightful observation that:

> Male readers' interest in romance comics no doubt partly stemmed from their provocative story titles and suggestive cover illustrations promising scenes of unbridled passion. Yet it is also possible that romance comics fulfilled an instructional role for adolescent boys, allowing them to gain some insight . . . about women's romantic expectations of men, the rituals of modern courtship and the 'acceptance' limits of (pre-marital) sexual desire – especially as such topics were rarely if ever covered in Australian men's magazines of the 1950s.

The readers of romance comics, as might be expected, were often young, low-level, white-collar workers. Harker (2009: 28) highlights suburban teenagers in Sydney reading the romance comics on their way to work: 'clerks, stenographers, bookkeepers or factory workers'.

However the conservative normative structure of romance comic narratives did not always follow the typical pattern as Patrick (2017: 241) shows:

> In 'Tomorrow for Tears', Joan Baxter falls in love with her handsome Latino dance instructor, Pedro Arbon, who persuades her to join him as his dancing partner at a nightclub in Panama. Joan happily agrees, but is horrified to learn they have been hired to perform at the seedy Café Oro, where she is expected to flirt with rowdy, drunken male clientele. Joan demands to know if Pedro ever intended marrying her, only to be told: 'Marriage wouldn't fit my temperament, Joan, it was never part of our . . . business arrangement, you know' (Powell, c.1956: 15). The story concludes with Joan returning to America, ashamed and repentant: 'I knew I had to atone for my recklessness . . . I had to win back my self-respect. And I knew it wasn't going to be easy' (Powell, c. 1956: 16).

The romance comic books were adapted to an Australian context by replacing American place name spelling and slang expressions with their Australian equivalent. Patrick (2017: 242) also indicates that Australian publishers would sometimes commission an Australian story for a romance comic:

> In 'Between Two Men', Lucy Collingridge is being pressurised by her lawyer and fiancé, Charles Flintridge, to sell Willaroonga Station to pay off her deceased father's debts. Distraught, Lucy returns to the family property, where she meets Garry Lorraine, a ruggedly handsome artist camping on the property's riverbank. They (inevitably) fall in love, but Lucy returns to Sydney tearfully explaining that she is betrothed to Charles. Garry follows her upon learning that Flintridge is attempting to swindle Lucy out

of her property which . . . contains valuable gold deposits. Garry exposes Charles's scheme and returns to Willaroonga Station with Lucy.

As Patrick observes, despite the change of locale, the melodramatic conventions still apply. In fact, there are of course strong parallels here with the romantic narrative of the Baz Lurhman film *Australia*, enlivened by two US-based celebrities who are Australians, Nicole Kidman and Hugh Jackman.

Conclusion

This chapter has analysed the interconnection between being young and in love. The chapter reviews the findings of the Pew Research Center's (2015) fascinating and detailed analysis of American teens 13–17. The report of the Pew Research Center shows how social media is integrally involved in the behaviour and practices of teens in their relational activity. There are positive and negative aspects to the report, and there are clear gender differences in the responses to the role of social media in relationships. A number of contemporary studies are reviewed in the chapter by Wilmot and Elley, confirming the gendered nature of relational behaviour, and showing that the outcomes are both positive and negative. The research by Bogle on 'hooking up' provides an overview of how relationships for young people have changed from their parents' generation. The chapter also looks at how, historically, the attitude of young people towards love and romance in Australia was impacted by the growth of romance comic books, and how many were adapted to suit an Australian context.

4

MODERN ROMANCE

Introduction

In this chapter, we focus on the characteristics of modern romance, and how contemporary relationships have changed from how relationships were established in the past. Chapter 4 focuses on the humorous exploration of contemporary romance by the US comedian Aziz Ansari (2015a) in his book *Modern Romance*. The book was written with the US sociologist Eric Klinenberg, and offers an interesting exploration of how people find love in contemporary society, particularly in the US. Issues such as meeting partners, the role of technology, and sexting, cheating and breaking up are all covered in Ansari's book. A wide range of issues are raised by Ansari which relate to intimacy and marriage, including the following: geographical proximity and partnership; average age of first marriage; choices in marriage patterns; online dating; how people met future partners; adultery and internet cheating; passionate and companionate love; and marriage and commitment. This chapter presents the socio-cultural analysis of love and intimacy outlined by Ansari and explores these issues. The Americanisation of love and intimacy is very evident in both sociological and psychosocial research on relationships as well as on the commercialisation of intimacy. The chapter reviews some of the influences on the commercialisation of love and intimacy in an Australian context. It also, finally, reflects on whether chick lit and romance novels offer new versions of romance or confirm traditional narratives of romance through the commercialisation of intimacy.

It should be noted that while Ansari's work is only one dimension of the research in this chapter, Ansari was accused in 2018 of sexual assault, and a summary of the allegations against Ansari can be found in Way (2018). A summary of Ansari's response to the allegations can be found in Harmon (2018).

History, geographical proximity and partnerships

Historically, Ansari (2015a) shows that people met and married not just in the same town, but often from the same street, or the same neighbourhood, block or building. In the past people travelled far less, and the scope of their lives were far more limited, thus relationships were more localised. James Brossard, a sociologist at the University of Pennsylvania, in 1932 looked through 5,000 marriage licences of people living in Philadelphia. One-third of the couples who got married had lived within a five block radius of each other, one out of six lived in the same block, and one in eight married couples lived in the same building.

Ansari cites the average age of marriage as being much later as is generally understood. In the United States, drawing on the Current Population Survey, Annual Social and Economic Supplements 1947 to 2014, the average age of marriage for men has risen to 29–30, and for women, 27–28. Ansari maintains that the reason for the later age of marriage include: pursuing education, moving through different jobs and developing themselves as people. Ansari says that this is called 'emerging adulthood'. Ansari (2015a: 6) says that: 'you didn't marry each other because you were madly in love; you married because you could make a family together. While some people said they were getting married for love, the pressure to get married and start a family was such that not every match could be a love match'.

Coontz (2006) in her book *Marriage, a History: How Love Conquered Marriage* argues that people did not get married because they loved each other, but expectations of love were different to those of today. As Coontz (2006: 7) states, passion was seen as extremely risky for getting married: 'Marriage was too vital an economic and political institution to be entered into solely on the basis of something as irrational as love'.

Ansari argues that things changed a lot in the 1960s and 1970s, including what people would expect to get out of marriage. Women's equality was a big driver of the way in which marriage was transformed. They went to college and university, they got good jobs, they achieved economic independence and they gained control over their bodies.

Choices in marriage patterns: online dating

Ansari sees the growth of social media as being a significant factor in the way relationships are established. This includes media like Twitter, Tinder and others expanding the scope of choosing partners. Ansari says that online dating has its origin in the 1960s with the emergence of the first computer dating service. They were not successful in the first instance because people did not have computers at home.

Classified advertisements were the way in which singles looked for partners in the 1980s and 1990s. This kind of advertising was not new; this form of advertising started in the 1690s, and by the eighteenth century, matrimonial advertising had become a flourishing part of the newspaper business (Cocks, 2009). The ads took off after the sexual revolution of the 1960s. Ansari says that weekly newspapers were full of ads for what are called 'thin markets', which include LGBT and older straight people and middle-aged divorcees.

Match.com was launched in 1995; it was different from other online dating sites and had some crucial innovations, instead of matching up people with an algorithm. Match.com let its clients select one another in real time. By 2005, Match.com had registered forty million people. Many other sites opened up as demand grew: eHarmony was for people looking for serious relationships, while Nerve was for hipsters. Most offered a quasi-scientific method of filtering through the use of algorithms.

Online dating and selecting a partner

The Pew Research Center's (2013) study *Online Dating and Relationships* by Smith and Duggan part of the Pew's Research Center's Internet and American Life Project found that one in ten Americans have used an online dating site or mobile dating app; 66 per cent of these online daters have gone on a date with someone they met through a dating site or app, and 23 per cent have met a spouse or long-term partner through these sites.

Smith and Duggan (Pew Research Center 2013: 9) maintain that:

> The rise of tech-enabled dating help has been one of the most striking developments of the digital era and these alternative ways of meeting and mating have risen at a time of fundamental change in the structure of marriage and divorce in America. The number of Americans getting married has been steadily declining, and today a record-low 51% of the public is currently married has been steadily declining and today a record-low 51% of the public is currently married . . . Americans are also waiting until later in life to get married, and other living arrangements -such as cohabitation, single person households, and single parenthood – have grown more common in recent decades. At the same time, marriage still holds wide appeal for those who have not tied the knot. Some 61% of men and women who have never married say they would like to get married eventually, while just 12% say they definitely do not want to marry.

The report usefully presents an overview of the differences between online dating and conventional offline dating, as follows (Pew Research Center, 2013: 9):

> Research into whether online dating actually produces more successful relationships or romantic outcomes than conventional (offline) dating is generally inconclusive although these sites clearly offer a qualitatively different experience compared to traditional dating. Some of these differences include: the ability to search from a deep pool of potential partners outside

of one's existing social networks; the ability to communicate online or via email prior to arranging for a face-to-face interaction; and matching algorithms that allow users to filter potential partners based on pre-existing criteria. . . . . Other research has indicated that the efficiency of online dating and the size of the potential dating pool compared with traditional methods make the process especially useful for people (such as gays and lesbians, or middle-aged heterosexuals) who may have limited options for meeting people [see Finkel et al., 2012].

In terms of a summary of the findings, Smith and Duggan (Pew Research Center, 2013) found that general public attitudes towards online dating have become much more positive in recent years, and social networking sites are now playing a prominent role when it comes to navigating and documenting romantic relationships. They found the following points: 11 per cent of American adults and 38 per cent of those who are currently single and looking for a partner have used online dating sites or mobile dating apps. The most common online dating sites used are Match.com, eHarmony and OKCupid.

Smith and Duggan (Pew Research Center, 2013: 2) maintain:

> online dating is most common among Americans in their mid-20s through mid-40s. Some 22% of 25–34 year olds and 17% of 35–44 year olds are online daters. Online dating is also relatively popular among the college-educated and urban and suburban residents . . . 38% of Americans who are single and actively looking for a partner have used online dating.

A second finding is that 66 per cent of online daters have gone on a date with someone they met through a dating site or app, and 23 per cent of online daters say they have met a spouse or long-term relationship partner through these sites. In addition, and moving beyond dates, one quarter of online daters (23 per cent) say they themselves have entered into a marriage or long-term relationship with someone they met through a dating site or app.

The report indicates that attitudes towards online dating are becoming more positive over time. They argue that 59 per cent of all internet users agree with the statement that 'online dating is a good way to meet people', a 15-point increase from the 44 per cent who said so in 2005. In addition, 53 per cent of internet users agree with the statement 'online dating allows people to find a better match for themselves because they can get to know a lot more people', a six-point increase from the 47 per cent who said so in 2005 (Pew Research Center, 2013: 2).

The report indicates that social networking sites offer a new online venue for navigating the world of dating and relationships. Today, six out of every ten Americans use social networking sites such as Facebook or Twitter, and these sites are often intertwined with the way they experience their past and present romantic relationships.

The following statistics on how Americans met their spouses between 2005–2012 show how online dating has changed how relationships are established:

- Online =34.95 per cent
- Work = 14.09 per cent
- Friends = 12.4 per cent
- School = 7.14 per cent
- Family = 4.4 per cent
- Bar or club = 5.68 per cent
- Place of Worship = 2.66 per cent
- Social Gathering =6.5 per cent
- Grew up Together = 4.92 per cent
- Blind Date = 1.73 per cent
- Other = 5.54 per cent

Ansari indicates that John Cacioppo et al. (2011b) from the University of Chicago, in a study between 2005–2012, says that one third of couples who got married in the US met through online dating. It was a bigger factor than other more traditional areas.

These findings are consistent with those of Rosenfeld and Thomas (2012), who have documented internet dating and the decline of other forms of connecting. In Cacioppo's survey 'How Couples Meet and Stay Together', he drew on a nationally representative study of 4,000 Americans, 75 per cent married or in a romantic relationship and 25 per cent single.

Comparing 1940 and 1990, in 1940, the most common way to meet a partner was through the family, and 21 per cent met through friends; about 12 per cent met through the church, and roughly the same proportion met in a bar, restaurant or work. By 1990, the family had become a less influential matchmaker, pairing up only 15 per cent of singles, as did the church, which had plummeted to 7 per cent. The most popular route to romance was through friends, 40 per cent. The proportion who met in bars was 20 per cent. By 2000, five years after Match.com was invented, 10 per cent had met on the Internet, and by 2010, nearly 25 per cent had. Now online dating has risen to 38 per cent of Americans (Pew Research Center, 2013).

Rosenfeld and Thomas (2012) in 'Searching for a Mate: the Rise of the Internet as a Social Intermediary' found that internet dating has changed the game even more dramatically in what they call 'thin markets': that is, people in same-sex relationships, but increasingly older and middle-aged straight people too.

As regards social stigma and online dating, there is still a social stigma attached for some people to online dating. Don Slater's history of online dating: *Love in the Time of Algorithms: What Technology Does to Meeting and Mating* found there was a huge discrepancy between what people say they wanted and who they matched up with.

Love in the time of algorithms

Eli Finkel et al., who published a paper in 2012 in *Psychological Science in the Public Interest*, argued that no algorithm can predict in advance whether two people will make a good couple. They say that: 'No compelling evidence supports *matching sites' claims that mathematical algorithms work*' (Finkel et al., 2012).

Finkel et al. (2012) set out:

> To understand how online dating fundamentally differs from conventional offline dating and the circumstances under which online dating promotes better romantic outcomes than conventional offline dating, we consider the three major services online dating sites offer: access, communication and matching. Access refers to users' exposure to and opportunity to evaluate potential romantic partners they are otherwise likely to encounter. Communication refers to users' opportunity to use various forms of computer-mediated communication (CMC) to interact with specific potential partners through the dating site before meeting face to face. Matching refers to a site's use of a mathematical algorithm to select potential partners for users.

Much of online dating, Finkel et al. say, is based on the faulty notion that the kind of information we can see in a profile is actually useful in determining whether that person would make a good partner. But because the kind of information that appears on a profile – occupation, income, income, religion, political views, favourite TV shows – is the only information, we over-value it. This causes us to make bad choices about who to go out with.

Finkel et al. come to the above conclusion based on the following pros and cons of online dating:

> In the final analysis, is online dating unique from and does it yield superior outcomes to conventional offline dating? The answer to the uniqueness question is an unqualified yes: Online dating is pervasive, and it has fundamentally altered both the romantic acquaintance process and the process of compatibility matching. The answer to the superiority is more qualified. Online dating offers access to potential partners whom people would be unlikely to meet through other avenues, and this access yields new romantic possibilities.
>
> On the other hand, the heavy emphasis on profile browsing at most dating sites has considerable downsides, and there is little reason to believe that current compatibility algorithms are especially effective.

Finkel et al. (2012) conclude as follows:

> Online dating functions best to the degree that it introduces people to potential partners they would have been unlikely to encounter otherwise

and facilitates a rapid transition to face-to-face interaction, where the two people can get a clearer sense of their romantic potential. As online dating evolves and matures, it seems likely that more and more of us will first encounter romantic partners online.

One example of such an online dating site was Tinder. Tinder was conceived by Rad and Mateen, two University of Southern California undergrads in 2011, who set out to create an online dating experience different from anything else. Like Facebook, Tinder's birthplace was college. But while Facebook began its rollout in the Ivy League, Tinder aimed for famous party schools like USC and UCLA. Within weeks, the app took off and was very popular on campuses. Within weeks, thousands of users had signed up, and 90 per cent of them were between the ages of 18–24.

Ansari talks about the role of privacy, cheating and the growth of the internet site Ashley Madison, which was designed to help people have affairs, with an estimated 11 million members in 2014. A growing number of people and a majority of young adults are more likely to break up with someone by text, instant message or social media than in person or by phone.

According to a 2014 survey of 2,712 18–30-year-olds, 56 per cent said they had broken up using digital media, with texting being the most popular method (25 per cent) followed closely by social media (20 per cent) and then email (11 per cent).

Passionate love and companionate love

Giddens has distinguished between passionate and companionate love (see Brooks, 2017). Some researchers have attempted to measure these concepts in different ways. Companionate love is neurologically different from passionate love. Passionate love always spikes early, then fades away, while companionate love is less intense but grows over time. Anthropologist Helen Fisher (2009), the author of the *Anatomy of Love*, was part of a research team that gathered and took brain scans of then-middle-aged people who'd been married an average of 21 years while they looked at a photograph of their spouse, and compared them with brain scans of younger people looking at their new partners. What they discovered is that: 'Among the older lovers, brain regions associated with anxiety were no longer active, instead there was activity in the areas associated with calmness'.

The transition from passionate love to companionate love can be tricky. In his book *That Happiness Hypothesis*, social psychologist Jonathan Haidt identifies two danger points in every romantic relationship: one is the apex of the passionate love phase, which then turns toxic, ending in divorce; second is the time when passionate love starts wearing off, and the conclusion is the person is no longer the right one.

Marriage and commitment

In recent decades, and in most developed nations, marriage rates have dropped precipitously. Philip Cohen (2013), one of the leading demographers of the family, has documented the steep and widespread decline in global marriages since 1970s. His figures show that 89 per cent of the global population lives in a country with a falling marriage rate.

In the United States, marriage rates are now at historic lows. In 1970, for instance, there were about 74 marriages for every 1,000 unmarried women in the population. By 2012 that had fallen to 31 per 1,000 single women, a drop of almost 60 per cent. In 1960, 68 per cent of all people in their twenties were married, compared with just 26 per cent in 2008.

Long-term cohabitation is on the rise. The rate of living alone has skyrocketed, particularly in cities such as Paris, Tokyo, Washington D.C. and Berlin. Marriages can lead to people living longer and being happier and healthier than single people. Good marriages can bring financial security, and better-off people are marrying more successfully than poor people, which increases inequality.

Consumption, romantic love and the Americanisation of love in Australia

The chapter has focused on many aspects of the Americanisation of romantic love and the practices of love contemporaneously in the US, and their implications for relationships globally. However, the Americanisation of patterns of romantic love have also been historically relevant. Teo (2006) has charted the Americanisation of Australian culture from the nineteenth and into the twentieth century. She has specifically focused on the culture of romantic love.

Teo (2006: 178) shows how the culture of romantic love was similar in the United States, Australia, Canada and Britain. However, she notes there were some distinct differences between Australia and the United States:

> Unlike nineteenth century American lovers who viewed romantic love as something highly mystical and mysterious, Australians generally tended to have more concrete and prosaic ideas about love. This was partly due to the fact that unlike American culture, romantic love was not sacralised in Australian culture.

Teo notes that the discourse around romantic love was characterised more by pragmatism in Australia than in the United States:

> The rhetoric of romantic love among Australians was never as intense, sublime or spiritualised as in the United States, neither was romance transformed into a new religion in Australia. Moreover, throughout the nineteenth and twentieth centuries, where the private correspondence among Australians

reveal an eloquence of emotional feelings, the public rhetoric of romantic love has been characterised by awkwardness, self-deprecation and even bathos, in stark contrast to public romantic rhetoric in the United States.

(Teo, 2006: 178)

Teo also highlights the more commodified approach to courtship resulting from the influence of the US, which is outlined elsewhere in this book (see Chapter 9). The particular focus of the Americanisation of romance was the institution of dating. As Teo (2006: 175) comments:

> Dating was controlled by men. . . . It was focused on consumption rather than production. . . . It was hedonistic in that pleasure was the goal, and pain was increasingly an unacceptable part of the experience of romantic love. And above all the same limited script of romantic consumption was widely broadcast and reinforced by advertising films, romance novels and romanticised commodities -especially what Eva Illouz has called 'ego expressive' commodities such as shampoo, perfume, deodorant and cosmetics [see also Chapter 9].

Teo also shows how Australian women's magazines often commissioned articles by American writers, including Rupert Hughes with his article 'What is True Love?', as well as significant women magazine writers such as Kathleen Norris. In addition, the number of articles on Hollywood relationships and divorces also increased. Whereas this was initially focused on women, Teo (2006: 191) also shows how this also extended to men:

> it was only after American magazines began to be imported to Australia in the postwar years, and the style of Australian advertising directed at men changed to a focus on them as consumers, that love letters from Australian men demonstrate the same notion of commodified romance that Australian women had become familiar with.

New versions of romance or a confirmation of traditional narratives of popular romance – chick lit and romance novels

Consumption of romantic love through a range of different forms of romantic novels and magazines, particularly driven by the US, has been shown throughout the chapters of the book. There has been much discussion about chick lit and whether it can be seen in the tradition of Mills and Boon or Harlequin, or whether these narratives offer a new version of romance. Some theorists argue that they offer a new version of romance which can be designated as postfeminist.

> Yvonne Tasker and Diane Negra (2005) . . . note parallels between chick lit novels and contemporary romantic comedy movies as post-feminist

> texts . . . . Usually the chick heroine, somewhat flawed in character, adopts a first person narrative style to tell a tale of self-transformation, self-realization, and self-empowerment (Georgina Isbister, 2008) as she navigates the messiness of her professional and personal life, including the quest for true love in the form of a suitable male partner (Imedla Whelehan, 2000).
>
> *(Rowntree, 2015: 508)*

In the first instance, the heroine's voice in chick lit is satirical as she struggles to negotiate conflicting feminine aspirations for non-heteronormative relationships (Isbister, 2008). Another element of chick lit which does give it some claim to postfeminist credentials is that this is not just a search for a love relationship, but the evolution of the self may be in other directions (Isbister, 2008; Mabry, 2006). Isbister (2008) says that the chick lit heroine may find satisfaction and empowerment through shopping, rather than in a conventional hetero relationship.

> According to Mabry (2006) a feature of the chick genre that distinguishes it from previous romance is its address to women as sexual players, rather than only as romantic players. Unlike her former counterpart, the chick heroine is portrayed as sexually active, and experiences rather than only anticipates sexual pleasure (Anna Kiernan, 2006).
>
> *(Rowntree, 2015: 509)*

In addition, the chick lit heroine has a number of sexual partners, at least until she finds love. Gill and Herdieckerhoff (2006) maintains that 'really good sex is reserved for Mr Right' (Rowntree, 2015: 509).

The chick lit heroine is overtly sexual, says Anderson (2000), but becomes more chaste when she meets the love of her life. Gill (2007) notes that 'women are represented paradoxically, with empowered and conventional sexual subjectivities, both states of which she is able to freely choose'(Rowntree, 2015: 509).

Linked to chick lit is the emergence of new-style romance novels and how feminist narratives inflect these new genres. In an article in *The Atlantic* by Jessica Luther (2013), entitled 'Beyond Bodice-Rippers: How Romance Novels Came to Embrace Feminism', Luther reviews a range of romance novels influenced by feminists and feminist thinking:

> Jackie Horne, author of *Romance Novels for Feminists* says that the women who now write romance novels grew up enjoying the benefits of the feminist movement. . . . In Alice Clayton's *Wallbanger* and Lauren Dane's *Lush* both heroines are adamant that their careers not suffer in order to make a relationship work. They negotiate long-term committed relationships with men who treat them as equals . . . these women seek out sexual pleasure and they enjoy sex.
>
> *(Luther, 2013)*

Another author, Cecilia Grant, is the author of historical romances such as *A Woman Entangled*. Grant makes the point that despite the feminist movement, the culture already prioritises relationships, and that 'finding a man' is the main goal. In this context, she argues that it is difficult to make the case for romance as 'a feminist-friendly medium'.

In Grant's first novel, *A Lady Awakened*, Grant sets out how feminist heroines may establish their own subjectivities in a relationship:

> the heroine uses the hero in order to get pregnant. She is not initially interested in emotional intimacy or love. The heroine is the one taking charge of her sexuality and her future, while it is the rake who we find crying about how he feels used and eventually begging his love for a long-time commitment.
>
> *(Luther, 2013)*

Luther shows how, for most of these writers, the most critical aspect for the heroines of the feminist romances is making choices, even though the choices available may be limited by convention or circumstances.

Thus, the key characteristics of heroines in 'feminist' novels is agency rather than transformation. Ana Cowan's message in her novel *My Lady Untamed* is that the heroines can make their own decisions, so feminist heroines are not offering alternative life patterns, they are simply choosing from traditional choices. This can be clearly seen in the novels of Victoria Dahl, as Dahl says: 'my characters are always feminists. Not in the declarative sense, but in the living-that-life-every-day sense' (cited in Luther, 2013).

Grant summarises the situation regarding the relationship between feminism and romance novels, as outlined by Luther (2013):

> Grant puts it succinctly, 'romance places where a woman is a *subject* of sex, rather than an *object*' . . . Grant echoes that romance novels have the ability to not only turn their readers on but, in fact, show their readers *what* exactly turns them on: 'romance (being written by women as it is) is a way for women to explore what we'd like sex to look like, and to define sexual success or validation for ourselves'.
>
> *(cited in Luther, 2013)*

Conclusion

This chapter focuses on 'modern romance' and looks at some of the characteristics of contemporary love and intimacy, particularly in relation to the US and its implications for love and intimacy globally. Some of the socio-cultural aspects of the concept is captured in Ansari's (2015a) *Modern Romance*. Some of the characteristics that Ansari develops include how people met and married, the average age of first marriage, the choices people make in future partners, online dating and 'cheating' as well as passionate and companionate love. Ansari's work is valuable in providing detail on patterns of online dating and in drawing comparisons between online dating and conventional dating. That chapter also follows valuable information on contemporary patterns of relationships through the Pew Research Center (2013) report on *Online Dating and Relationships*, which contextualises online dating in the context of other traditional mechanisms for meeting partners. Ansari also focuses on 'love in the time of algorithms', and Finkel et al.'s (2012) work highlights the limitations of online dating in the selection of partners. In addition, Ansari shows that not only are relationships established online, but they are also brutally ended online. The Americanisation of love and intimacy is very evident in both sociological and psychosocial research on relationships, as well as on the commercialisation of intimacy. The chapter reviews some of the influences on the commercialisation of love and intimacy in an Australian context. It also considers chick lit and recent romance novels and asks if they offer alternative romance narratives or confirm traditional narratives of romance through the commercialisation of intimacy.

5

LOVE IN THE MOVIES

Introduction

This chapter focuses on how love and intimacy have been represented in films. I look at the socio-historical and socio-cultural context of love and intimacy, and the way romance has been represented historically with the emergence of different genres, including romance and film noir, romantic comedies and bromances ('homme-coms'). I also look at the way in which the '#MeToo Movement' and the significant reporting of Ronan Farrow (2017, 2018) in *The New Yorker* has impacted women in the filmic and television industries. I look at heteronormative perspectives on love in the movies, and examine movies where challenges are made to the heteronormative dominance of movies. The chapter also looks at the psychosocial context of masculinity and jealousy. Yates (2007) explores the psychosocial and socio-cultural construction of masculinity and jealousy and how it is represented in film. Stearns (1989) tracks the history of jealousy in American history. Mullen (2018) argues that whereas jealousy was once an accepted public emotion, male jealousy was linked to a male sense of entitlement. Mullen (2010) also notes that historically, this changed in the twentieth century with the emergence of feminism, and male jealousy is seen in a different way and linked to domestic violence and stalking. This chapter also explores a number of cinematic theories which examine a psychosocial approach to understanding emotions in film, including the work of Laura Mulvey (1975; Creed, 1993; Mulvey, 2015).

Changing relationship structures after the social revolution of the 1960s, including the breakdown of traditional marriage structures, the increase in divorce rates and the growth of the women's movement, resulted in different representations of love, intimacy, sex and relationships in films. Shumway (2003) maintains that by the 1970s, sex had officially become part of courtship rather than marriage, and this was projected through the adjustment of Hollywood romantic comedies. Neale (2009) describes 1970s romantic comedies as the phase of 'nervous romance'. Films started to reflect a less harmonious picture of relationships and far more volatility. Shumway maintains that this enabled viewers to relate more fully with the picture being presented on the screen. An example of this is the film *Annie Hall* (1977), directed by Woody Allen, who presented a humorous, intellectual and high-end view of how relationships were changing. Woody Allen stars opposite Diane Keaton in the film and he falls in love with her character, but she struggles to maintain intimacy without marijuana. The film ends with the failure of the relationship as she ends it, thus moving away from traditional endings, empowering female needs for intimacy and representing a more realistic frame of reference for the needs of the modern independent woman. Alberti (2013: 160) notes:

> *Annie Hall* (Woody Allen, 1977) represents the exemplary turning point in this trend, a narrative that abandons the culturally dominant goal of the stable long-term heterosexual romance. . . . The last shot in the movie signals this abandonment, with the camera focused on a busy New York city street after both lovers leave the frame walking in different directions, away from each other and away from narrative closure that historically marked the genre.

In the 1980s and 1990s, relationships and family patterns were transitioning from a single overriding norm of heterosexual nuclear relationships, to a sexually and culturally diverse range of romance, consisting of minority ethnic groups and LGBTQ groups (Giddens, 1992; Giddens and Sutton, 2013). These films took a variety of different forms. These include the emergence of the 'bromance' (homme-com) and a different type of masculinity, as shown in films such as *Brokeback Mountain* (2005), and the emergence of the independent woman, as shown in films such as *Thelma and Louise* (1991), *Sex and the City* (Michael Patrick King, 2008) and *Carol* (2015). Other films focused on the emotions emerging from the changing nature of relationships which produced adultery and jealousy, as in *Fatal Attraction* (Adrian Lyne, 1987) and *Unfaithful* (Adrian Lyne, 2002).

Breaking down of heteronormativity: bromances and the [re]construction of masculinity in romantic comedy

In his article '"I Love You, Man": Bromances, the Construction of Masculinity, and the Continuing Evolution of the Romantic Comedy', Alberti (2013) explores:

> the contemporary sub-genre of the bromance-ostensible romantic comedies centered on confused homosocial/homoerotic relationships between putatively straight male characters-as examples of this, generic exploration of 'other types of relationships' not defined by the conventional codes of the

heteronormative romantic comedy explorations driven by and responding to the 'new climate of social and sexual equality between men and women'.

The term 'homme-com' was introduced by Jeffers McDonald (2009) and includes texts or film narratives which were slanted at males and provided an alternative to traditional 'chick flicks', which appealed to a female audience. Claire Mortimer (2010: 135–136) describes bromance as follows:

> The bromance is an ironic take on the romantic comedy, which can appeal to both genders at the box office, reaching out to the male audience that would regard the romantic comedy as a 'chick-flick'. These films work to reclaim masculinity for a generation that sees feminism as a historical movement and is familiar with conflicting representations of men in popular culture, ranging from the metrosexual icon of David Beckham to the macho posturings of many hip pop stars.

Homme-com also incorporates the friendship of bromance between male friends instead of just a traditional relationship. The framework of bromance offers greater opportunity for men to discuss issues that traditional macho men wouldn't need to or be inclined to. Despite this, as Alberti shows, even more 'progressive' movies have aspects of neo-traditional romance and include alpha-males.

Alberti shows that the *Sex and the City* franchise represents a kind of parallel development to the bromance. The original series on HBO was ground-breaking in terms of the issues around sex and gender that the television series raised. The films were more conventional, with a neo-traditional narrative, but the television series raised challenges to conventional relationships on a number of different levels. Some of the contradictions raised by the series include consumer and sexual desire, romance, pleasure and marriage. Alberti (2013: 164) comments that as regards models of masculinity:

> in *Sex and the City*, the character of Mr Big foregrounds the degree to which this conventional Alpha male character disrupts the emerging algebra of the post-patriarchal romantic comedy, as a core topic among the four lead women in the show focuses on whether a long-term relationship with such a throwback construction of masculinity (echoed in the form of Don Draper from *Mad Men* . . .).

As mentioned above, the film versions of *Sex and the City* were more typical of neo-traditional romantic narratives than the film series. Alberti (2013: 164), drawing on Mortimer (2010), highlights some of the reasons why Mr Big is characterised in the way that he is in the film and series:

> in discussing the first movie version of *Sex and the City* (Michael Patrick King, 2008), Claire Mortimer points out that almost vestigial role that Mr

Big plays in the movie: 'the rather dull figure of Mr Big is unconvincing and almost redundant in comparison to the compelling female characters' (Mortimer, 2010: 39). This redundancy – echoed in the character's ironic name –suggests that the Alpha male is a mere placeholder in the movie, a part of generic algebra that struggles to maintain relevance even as the woman characters in *Sex and the City* register the same generic self-consciousness we see at operation in the bromances.

Alberti also points out that the characters of Mr Big and Don Draper, as well as the actors who portray them (Chris Noth and John Hamm), physically present the characteristics of Alpha masculinity. Alberti (2013: 171) comments that: '[b]oth as well are meant to register as the way leading men used to look in the case of Mr Big, anachronistically, and in Don Draper as the representation of an extinct species.'

The idea of challenges to heteronormative patterns has not always been humorous or ended romantically. For example, in the film *Rebel Without a Cause* (1955), the James Dean character who was designated as the 'rebel' actually fitted the traditional model of masculinity of the time. However, the Sal Mineo character, who is portrayed as being attracted to Dean's character and who is seen as latently gay, is shot dead in the film.

Similarly, women who rebel against conventional patterns also met with typically negative fates. In *Thelma and Louise* (1991), Susan Sarandon and Geena Davis show strong feminist bonding and disdain for masculine dominance, and again die at the end of the film in driving their car over a cliff and into the Grand Canyon.

While there is a rise in the biopics of gay rights icons within mainstream cinema, there is still a double standard between same-sex relationships and how they are seen. An example of this can be seen in the way *Brokeback Mountain* (Ang Lee, 2005), featuring a romance between two cowboys in 1950s America, was received, and *Carol* (Todd Haynes, 2015) (ironically produced by The Weinstein Company) featuring a romance between two women in 1950s America which was ignored by the Academy in terms of an Oscar.

Brokeback Mountain concentrates on two men who fall in love when they are young, and then continue a twenty-year affair while living conventional lives with wives and children. One of them is killed in a possible hate-crime, but this is never clarified in the film. *Carol* focuses on the relationship between two women, one older with a husband and child (although she is clearly gay and her marriage is breaking up), and a younger woman who feels trapped in a long-term relationship. In this case, neither of the women die and it is unclear whether they will end up together, although that is the indication. As stated, *Brokeback Mountain* was given nominations in the Academy Awards in the Best Picture and Best Director categories, whereas *Carol* was not. It has been noted by Thompson (2014) that these differences were due to the demographics of members of the Academy.

Masculinity, jealousy and change

The role of jealousy is a significant one in understanding love and intimacy in film. Yates (2007) examines the relationship between jealousy and masculinity, and explores a psychosocial perspective on jealousy. Both Mullen (1991a) and Stearns (1989) explore the history of jealousy in Western society and explore changing social and cultural attitudes to understanding jealousy. Stearns argues that whereas jealousy was once an accepted public emotion, it was also linked to gender norms and relations of possession. Male jealousy was linked to a sense of male entitlement and of protecting what they saw as their property. This changed in the twentieth century when feminists challenged the sexual double standard and gained more social, economic and legal rights. Male jealousy is more widely linked to domestic violence and stalking in contemporary relationships (see Mullen, 2010a).

Psychosocial and sociological perspectives have examined the relationship between emotions and social change. Elias (2000) shows that public displays of anger such as jealousy were perceived as being antisocial, as he developed in his thesis of *The Civilising Process* (see Brooks, 2014a; Lemmings and Brooks, 2014). Stearns (1989) draws on Elias in his discussion of the changing 'emotionality' of jealousy, and discusses the ways in which jealousy, which is linked to attachment and the need for another person, is often considered in a society underpinned by individualism and the values of a free market, whereas Yates (2016) notes 'self-sufficiency, narcissism and self -promotion are the goals'.

Pathological interpretations of jealousy

Films still represent a double standard when it comes to how jealousy is portrayed. This can be seen in two films directed by the same film director, Adrian Lyne; these are *Unfaithful* (2002) and *Fatal Attraction* (1987). Simply put, there is far greater sympathy for the murderous Edward (Richard Gere) in *Unfaithful* than there is for the jealous slighted lover Alex (Glenn Close) in *Fatal Attraction*. This could have to do with fairly superficial explanations, such as the actors playing the roles and the ideology being transmitted by the director.

The film *Unfaithful* is analysed by Yates (2007). In the case of *Unfaithful*, the film explores 'the themes of sexual fidelity and the fantasy who may, as with Edward in *Unfaithful* find himself at the mercy of forces beyond his emotional control' (Yates, 2007: 142). A number of key themes in these films are surfaced by Yates (2007: 147): 'a recurring theme of these films is the danger posed to the family by adultery, then another is the danger of female sexuality, and also, as in *Fatal Attraction*, the fearful and pathological nature of female jealousy'.

The role of Gere and his filmic credibility is drawn on in *Unfaithful*. As Yates (2007: 149) comments:

> Gere's performance as Edward, the cuckolded husband and father, in *Unfaithful*, should be seen in the light of recent anxieties about the fragmentation

of the family in Europe and post-Clinton America . . . social, political and cultural changes have contributed to what some have defined as a Western 'crisis of masculinity'. . . . *Unfaithful* is a film about the precariousness of the nuclear family and the instability of gendered family relations and parenthood within the late modern world as seen from the male point of view.

The role of the director Adrian Lyne is also significant in the ideology conveyed in these films. Lyne has been accused of misogyny for his cliched portrayal of women in 'raunchy sex scenes' in a range of films in the 1980s and early 1990s, including *91/2 Weeks* (1986), *Fatal Attraction* (1987) and *Indecent Proposal* (1993). As Yates (2007: 143) comments:

> Critics of Lyne point to the non-too-subtle way that his films preach a certain kind of reactionary morality that reinforces the myth of the American dream embodied through the perfect image of the happy nuclear family (Craving, 2002: 138). The moral message of these films arguably serves as a warning that adultery and the betrayal of one's family come at a price.

The sympathy of the audience is drawn towards Edward, despite his brutal murder of his wife's lover. However, there is no sympathy encouraged for Glenn Close (Alex) as the mistress in *Fatal Attraction*. As Yates (2007: 143) comments:

> In *Fatal Attraction*, the price of the husband's adultery was the murderous jealousy of the unhinged mistress; in *Indecent Proposal*, the price of the wife's infidelity for money was literally one million dollars and the potential destruction of a marriage; and in *Unfaithful*, the wife's infidelity leads to the husband taking jealous revenge with disastrous consequences.

Leonard (2009: 10), in her book *Fatal Attraction*, discusses how Lyne researched the role of the single independent woman to represent her in the film as follows:

> The presentation of Alex's apartment as playing host to little more than empty air was no doubt deliberate, and, while publicizing the film, director Adrian Lyne, made a habit of repeating the fact that in order to get a feel for how single woman's apartment should look, he previewed Polaroids of actual editors apartments in New York. His remarks on what we saw in those shots did nothing to endear him to feminists, for he was particularly fond of pointing out that he found such spaces rather spartan and depressing. In one of the many reported iterations of his impressions he notes in the director's commentary that all the women had were 'piles of manuscripts by the bed; it was rather mournful.' When associated with Alex, white colludes with a larger visual project

of recording a life that is harsh and empty, an existence she tries hard to populate. . . . Connoting not abundance but absence, Alex's white life has a harsh antiseptic feel.

Masculinity in crisis

The point made by Yates (2007) goes beyond the issues emerging from the representation of masculinity; she raises the concept of 'masculinity in crisis' and 'the broader unravelling of masculinity as a social and cultural construction'(Yates, 2016). Yates also comments that 'jealousy is a useful concept to explore such issues because it tests the capacity to cope with complex emotions that arise in relation to feelings of wounded narcissism and anxieties about separation, difference, rejection and loss'.

Yates (2007, 2016) raises the question of whether such representations show a more reflexive and fluid forms of masculinity, or, 'more negatively, are such representations at the expense of the female protagonist and so represent a defensive shoring up of hegemonic masculinity in a new guise?' She argues: 'when faced with the spectacle of male sexual jealousy, with all its cultural connotations of cuckoldry, shame and humiliation the viewer may find it hard to watch such images and wish for certainty and a fantasy of male mastery to be restored'. Yates (2016) also maintains that today, flirtation is deployed in order to defend against the risks of attachment and the shame of jealousy.

Masculinity and romantic love in Australian films

One of the most distinctive aspects of Australian film is the representation of masculinity, and film has been a significant vehicle in Australia for conveying a number of significant issues regarding love, intimacy and gay relationships. Australian film is not really nuanced, but quite unambiguous in the messages being conveyed, whether it is about heterosexual marital relationships or same-sex marriage.

One aspect of this is captured by Nicholls (2014) in his analysis of male melancholia in marital relations in the Australian film of 1997, *Thank God He Met Lizzie*, with internationally acclaimed actors Cate Blanchett and Richard Roxburgh. The hollowness of the bourgeois marriage of Sydney has many parallels with the humour and despair of the Woody Allen genre of movies in the US. What is perhaps more interesting than the film itself is its location within the evolution of Australian film in the 1990s, in terms of how it represents romantic love. As Teo (2014a) comments:

> The 1990s opened with an exuberant celebration of Australian multicultural love in Baz Luhrmann's *Strictly Ballroom* (1992), but increasingly, this was a decade when Australian culture seemed to exhibit a loss of confidence

in narratives of romantic love dominated as it was by two other internationally popular Australian films where friendship was more enduring than romantic love: *Muriel's Wedding* (1994) and *The Adventures of Priscilla, Queen of the Desert* (1994). *Perhaps Thank God He Met Lizzie* fits into this cultural melancholia about love.

Rayner (2017), in his article 'Romantic Love in Australian Cinema', provides an interesting summary of romantic love in Australian film, which he argues often capture traditional melodramatic conventions. Enker (1994: 218, 220) locates Australian film within a broader international cinematic context:

> Examining romantic relationships between men and women in Australian films is illuminating if only as a guide to the filmmakers' collective discomfort with heterosexual love stories and scepticism [sic] about the possibility of enduring passion. There is a striking absence of the grand passions that are intrinsic to, and characteristic of, French and American cinema [. . .] perhaps this is why Australian films have repeatedly examined bonds between men and women. Certainly there are very few films that depict a romantic relationship between the first flutter of attraction to a blissfully happy ending for the couple [. . .]. Australian cinema seems sceptical about the capacity of love and particularly passion to endure. And even when it flickers for a while, it generally dies.

Baz Luhrmann's films are clearly famous for their dramatic and stunning visual presentation, and they are clearly not sophisticated characterisations of love, but they form a genre in their own right. As Rayner (2017: 174) comments:

> Luhrmann's films offer romantic love not simply as a feature of narrative but as a function of cinematic spectacle in which excessive romantic yearnings are positioned as comedic (*Strictly Ballroom*, 1992), tragic (William Shakespeare's *Romeo and Juliet*, 1996), tragi-comic (*Moulin Rouge*, 2001), and epic (*Australia*, 2008) genre framework.

Perhaps the film which paid deference to romantic conventions most fully was Luhrmann's *Australia*. There are many dimensions to the film, including the character of Australian masculinity captured by two of the foremost Australian character actors, Bryan Brown and Hugh Jackman; the landscape of Australia and the relationship of the characters to the land; the parallels between American epic cinema and that captured in the epic nature of Australia; and, Rayner also adds, 'a self-reflective play with narrative expectations and generic convention' (Rayner, 2017: 190). Rayner provides an interesting synopsis of the film:

> The cliched romance between Lady Sarah (Nicole Kidman) and the Drover (Hugh Jackman) might appear risible in its visual floridity. At the

ball in Darwin, Sarah's capitulation to King Carney (Bryan Brown) is forestalled by the Drover's entrance clean-shaven for the first time in the film, his matinee idol impact is signalled by his arrival in blurred long-shot being replaced by a rose-tinted full-face close-up . . . . To polite society, the Drover 'is as good as Black': *Australia*'s visual resemblance to an Antipodean *Gone with the Wind* (1939) belies its amalgamation of a cloying (White) Australian romance with the admission of countless unacknowledged, consensual and non-consensual inter-racial relationships.

Australian film has also been used to make significant political points, including the relationship of masculinity and same-sex love and marriage. Boucher and Reynolds (2017) in 'Same-Sex Love in Late Modern Australia: On the Political Straight and Narrow', provide a fascinating analysis of Australian 'mateship' and a loving gay relationship, in the film *The Sum of Us*. Boucher and Reynolds (2017: 342) summarise the key issues:

> *The Sum of Us* told the bittersweet story of a pair of 'red-blooded Aussie blokes' whose father-son bond was characterised by the laconic care and irreverent humour of mateship: working class widower Harry (played by Jack Thompson) and his plumber, rugby playing, beer-drinking gay son Jeff (played by Russell Crowe). Mateship had, of course, long been mythologised as a rational virtue in Australian cultural production . . . . The only difference between Jeff and his filmic antecedents was that when '[Jeff] goes to the pub, he hoped to strike up a friendship and score with a bloke, not a sheila' (McDonald, 2009).

What is interesting about the film is that, as Boucher and Reynolds note, 'homosexual desire thus made a determined entrance into the homo-social world of Australian mateship on screen', and so early on in the general social narrative towards same-sex relationships and marriages. In addition, the search for romantic satisfaction by both Jeff and Harry in their different spheres is undermined by the fact that their partners are unable to cope with a situation where 'homosexuality is normalised'.

Politically, as Boucher and Reynolds show, this was a pivotal point in the democratisation of romantic love in Australia, and established gay men's relationships as part of a legitimate national narrative. In fact, they note that both *The Sum of Us* and *Priscilla, Queen of the Desert*, both released in 1994, were 'making a bid for acceptance and legitimacy in a social and cultural context where homosexual men and women faced significant discrimination' (Boucher and Reynolds, 2017: 344).

Psychosocial approaches to love in the movies

Psychoanalysis has historically played an important part in the analysis of film. Within film theory, one of the most significant figures is Laura Mulvey (1975;

Mulvey et al., 2015; Mulvey and Backman Rogers, 2015). In her iconic article on the psychosocial analysis of film, 'Visual Pleasure and Narrative Cinema', Mulvey (1975), argues that: 'Psychoanalysis as a discipline supplied us with a conceptual vocabulary to scrutinise questions of sexuality, gender and social interpellation via the regulation or control of sexuality'(2015: 68–69).

In Mulvey et al. (2015), van den Oever comments that 'Visual Pleasure and Narrative Cinema' was derived from psychoanalytic theory and the concept of a patriarchal order, combined with feminist theory that framed the perspective. The application of the model was, for Mulvey, Hollywood films and to major directors like Hitchcock. As Mulvey (2015: 74) comments:

> Hitchcock's films work very well within such a psychoanalytic paradigm and, as such, exemplify how sadism, voyeurism, and fetishism can be embedded within the situation of anaemic spectatorship. Of course, with the advent of digital technology and the alteration of how we watch, analyse and consume films, our relationship to these kinds of images has changed.

Mulvey's shows how the dominance of the 'male gaze' reinforces male narcissism and patriarchal masculinity. Her work draws on Lacanian perspectives in showing that when we watch a film or enter a cinema we move into a dream-like state, and awaken infantile fantasies (Lacan, 1977). We become absorbed by the stories and figures on the screen. Lacan relates cinematic viewing to 'when a child first recognises its own image in the mirror' as they recognise the picture and relate it to the actual, giving an illusion of reality. Other theorists, such as Creed (1993), 'apply the psychoanalytic ideas of Karen Horney and Melanie Klein to film in order to discuss phantasies of the all-powerful mother and the ways in which these are evoked in film' (Yates, 2016). These are illustrated in films such as *Fatal Attraction*.

Mulvey's view of the 'male gaze' sees Hollywood cinema based upon a tension between narrative and spectacle, where men have control over the action, whereas women are positioned as objects of the male gaze. This is achieved through a series of three (male) looks which include: the look of the camera; the look of the male actor; the look of the audience. Mulvey's views have of course been challenged, and the question raised of what are the implications for the 'female gaze'. As Yates (2016) notes: 'women can be active heroines and viewers may identify with *the active male subject*. Men are not always subjects of the active gaze, they may also be objects of the gaze'.

The '#MeToo Movement' and the fall of sexually abusive men in film and television

A new and darker side of the film and media industries has been exposed recently through the work of Ronan Farrow (2017, 2018) in *The New Yorker* over the period from 2017 to 2018. While other sexually abusive men were exposed, including Roger Ailes and Bill O'Reilly at Fox, Matt Lauer at NBC and Charlie Rose at CBS, the focus of Farrow's work in exposing the sexually abusive Harvey Weinstein and CBS's Les Moonves and Jeff Fager has done much to advance the '#MeToo Movement'. All have now been sacked and their careers ended.

Farrow's (2017) article 'From Aggressive Overtures to Sexual Assault: Harvey Weinstein's Accusers tell their Stories' in *The New Yorker* built on the reporting in *The New York Times* where two journalists revealed multiple allegations of sexual harassment against Weinstein, an article that led to the firing of four members of the Weinstein Company's all-male board and Weinstein himself. As Farrow (2017) commented:

> Four actresses, including Miro Sorvino and Rosanna Arquette, told me they suspected that after they rejected Weinstein's advances or complained about them to company representatives, Weinstein had them removed from projects or dissuaded people from hiring them.

Weinstein's demise is, of course, well documented, and his career ended as well as his marriage. A similar pattern of sexually abusive and aggressive behaviour is seen in the behaviour of Les Moonves at the top of CBS. Farrow's (2018) article exposing Moonves and implicating his colleague Jeff Fager, chairman of CBS News and executive producer of '60 Minutes', has resulted in the sacking of Moonves and the very recent sacking of Fager (13 September 2018).

Ironically, as Farrow (2018) shows:

> Moonves had become a prominent voice in Hollywood's #MeToo movement. In December, he helped found the Commission on Eliminating Sexual Harassment and Advancing Equality in the Workplace, which is chaired by Anita Hill. 'It's a watershed movement', Moonves said at a conference in November. 'I think it's important that a company's culture will not allow for this. And that's the thing that's far reaching. There's a lot we're learning.'

Farrow (2018) points out that the real sexually abusive behaviour of Moonves had a significant impact on the culture of CBS as a whole:

> Thirty current and former employees of CBS told me that such behaviour extended from Moonves to important parts of the corporation, including CBS News and '60 Minutes' . . . During Moonves's tenure, men at CBS News who were accused of sexual misconduct were promoted, even as the company paid settlements to women with complaints. It isn't clear whether Moonves himself knew of the allegations, but he has a reputation for being closely involved in management decisions across the network. Some of

the allegations, such as those against the former anchor Charlie Rose, as reported by the Washington Post have already become public.

Some of the women who spoke to Farrow who experienced sexual assaults from Moonves included Illeana Douglas, Janet Jones, Christine Peters and Dinah Kirgo.

The abusive behaviour by Moonves and Rose, which was known about by Fager, created an environment where the abuse of women was seen as routine. As shocking as this may seem to those outside the industries, it was widely known about and feared by those inside. As Farrow (2018) comments:

> One former producer at CBS described CBS News as a 'frat house' . . . She added, 'Fager seemed to encourage that climate' . . . Katie Couric, who was an anchor at the network and a contributing correspondent for '60 Minutes' from 2006 to 2011 when Fager helped force her out, told me that it 'felt like a boy's club, where a number of talented women seemed to be marginalized and undervalued.'

As I complete this book, Fager has been sacked from CBS, ostensibly for threatening a journalist to not include material that was detrimental about him in her writing; her response was to publish the emails from him. Fager was sacked for behaving in a way inappropriate to CBS values. As mentioned, CBS is not alone, and Fox and NBC have both gotten rid of high-level powerful men who undermined the brand and abused their position, including Roger Ailes, the chairman and CEO of Fox News.

Conclusion

Chapter 5 brings together a range of different perspectives and concepts framing the analysis of love and intimacy in filmic and televisual discourses. The chapter reviews a range of genres, including romantic comedies, film noir and 'bromances' (homme-coms), as well as a range of movies reflecting psychosocial perspectives around masculinity and jealousy. A socio-historical and socio-cultural analysis reviews a range of ground-breaking films which chart the changing nature of love and intimacy in a wider social context. These include *Annie Hall*, *Thelma and Louise*, *Sex and the City*, *Brokeback Mountain*, *Carol*, *Fatal Attraction* and *Unfaithful*. Many aspects of love, intimacy, sexuality and jealousy have been considered in this chapter. In addition, the chapter reviews the changing nature of masculinity and jealousy and its representation in psychosocial dramas such as *Unfaithful*. This chapter focuses on the classical contribution of Laura Mulvey and her ground-breaking analysis of the 'male gaze'. Finally, the chapter has reviewed the dark side of film and television industries through the widespread sexual abuse of women in the industry, and the way the #Me Too movement has exposed sexual abuse by powerful men in the film and television industries.

Now the exposure of these men, including Harvey Weinstein, Les Moonves, Jeff Fager, Roger Ailes, Matt Lauer, Charlie Rose and Bill O'Neill (among others yet to be exposed), has destroyed their careers and made them unemployable, and has created a different and intolerant environment in these industries where such behaviour will be extinguished, and the perpetrators humiliated and socially ostracised.

6

LOVE AND INTIMACY IN MARRIAGE

Introduction

This chapter explores historical and cultural perspectives on marriage and traces different types of marriage in different cultural contexts. Coontz (2016) highlights the fact that traditionally, marriage benefited men rather than women. Additionally, Coontz (2015) looks at marriage and social change in the US. This chapter also explores the emergence of same-sex marriage in the UK and the US. An additional area explored in this chapter is the area of migration, marriage and intimacy though the work of Beck and Beck-Gernsheim (2014), which looks at 'commercial matchmaking' and mixed-nationality marriages, as well as women who migrate in search of marriage. Constable (2005a, 2005b, 2006) has written extensively on the role of marriage and migration and explores what she calls the 'cultural logic of desire'. The chapter also examines the issue of 'mail-order brides'. This chapter looks at heterosexual and same-sex marriage and considers whether marriage is still an important factor in relationships. The chapter also focuses on the way in which traditional narratives of love and romance, drawing on an American context within romance novels, have an impact in a post-colonial context like Australia.

Historical and cultural perspectives on marriage

The primary function of marriage historically was, of course, to expand the family labour force and to make alliances (see Brooks, 2017; Coontz, 2015, 2016). Illegitimate children were called 'love children' and, as Coontz (2016) comments, this was somewhat ironic, as of course love was considered a threat to the main function of marriage. Coontz (2016) notes that in mid-nineteenth century European cities, thousands of illegitimate children died each year. Coontz also notes that historically, the idea of marriage based on 'nobility and dignity' was almost exclusively for the husband, who had complete legal rights over the wife and children. She argues that, far from protecting wives, marriage deprived them of all rights, including rights against sexual assault.

On June 28, 2015, the United States Supreme Court ruled five to four that marriage was a fundamental right and could not be denied to gays and lesbians. John Roberts was the Chief Justice who dissented, and who claimed that 'marriage was a union of a man and a woman'.

Stephanie Coontz (2016), as a sociologist, shows that John Roberts's perspective reflects a limited Western perspective on marriage. She disagrees with him, and says that Mosuo or Na of China existed for thousands of years without formalised marriage. In addition, she says that while most marriages are heterosexual historically, the most common form of marriage was not one man and one woman, but one man and several women, as seen in a number of polygamous cultures.

The Na of China are farmers in the Himalayan region who live without the institution of marriage. Na brothers and sisters live together their entire lives, raising the women's children. Like other societies, the Na respect the incest prohibition; they practice a system of furtive or inconspicuous night visits, during which a man goes to a woman's home.

Coontz uses these examples to illustrate that there is no universality of marriage, as some societies around the world have functioned successfully without the practice of marriage, such as the Mosuo of China and their practice of 'walking marriages'. In this culture, it is customary for the men to walk and visit their 'partner' during the night and return home next day. Men and women lovers live separate lives, and never cohabitate. Anthropologists deem this community 'serial monogamists'. Within the Mosuo, because neither sons and daughters leave home, there is no preferential treatment for any gender, with focus instead on maintaining gender balance within a household (Lugu Lake Mosuo Cultural Development Association, 2006; National Geographic, 2016).

Marriage and social change in the US and the UK

In her analysis of the US, Coontz (2015) looks at the kind of changes which characterise changes in married life in the US. One of the key changes has been numbers of women working full-time in the workforce. She comments as follows:

> In 1963, less than 20% of American children younger than age 6 had mothers in the labour force. Today 65% of children younger than age 6 have a mother who is working or seeking work. In 1960, only one third of women were still employed 5 years after graduating from school. Today women are increasingly permanent, lifelong members of the labor force. In fact, women are the sole or primary earner in 4 out of 10 American families.

Coontz (2015) also talks about the changing structural framework of marriage and the family in the US:

> Fifty-three years ago, the median age of marriage for women was just over 20 years younger than today. Of adults age 25–29, 70% were married (and 90% of women in that age group) compared to only 25% today. Only 13% of the households contained just one person, most of them, elders who had lost a spouse. In 1960 people simply didn't live alone for any significant amount of time unless they were widows or widowers. Nor did they enter into long-term commitments, personal or financial, outside of marriage.

Coontz (2015) maintains that unmarried couples accounted for 1 per cent of all households in 1960. Just one child in twenty was born to an unmarried woman, and there were only 35 divorced people for every 1,000 married people. She goes on to say that today 40 per cent of children are born out of wedlock, and there are approximately 175 divorced individuals per 1,000 married ones. In 1980, 4 per cent of individuals age 65 and older were divorced. By 2008, that had tripled, transforming the role of marriage in old age as well as in young adulthood.

In addition, Coontz comments that more same-sex couples are openly raising children. For example, 26 per cent of same-sex couples in the Southern US have children in the home, 24 per cent in New England, and 21 per cent in the Pacific states. She also says that we are seeing an eclipse of the male-breadwinner households. On the issue of divorce, in the 1950s, people who married later than average had a greater chance of divorce. Today, every year a woman delays marriage up to the age of 35 decreases the chance of divorce. After age 35, the risk doesn't continue to go down, but it does not go up.

Until the 1950s, highly educated women were less likely to marry than less educated women. But for women born since 1960, college graduates and women with higher earnings are now more likely to marry – and much less likely to divorce – than women with less education and lower earning power.

In the UK, a survey of divorcing couples during the post-war period found that violence, adultery and bad habits were the key factors in women asking for divorce. Today, women are more likely to complain about long working hours, lack of communication and failure to participate in household chores. In the UK, a recent study found that a man's failure to do housework determined whether or not the wife was employed and was a strong predictor of divorce. Both the failure to engage in housework and the implications of the opportunity of the wife to be employed are factors in the greater propensity of divorce. For most of the twentieth century, couples who lived together before marriage had a greater chance of divorce than those who entered directly into marriage. However, since the mid-1990s, cohabitation before marriage is not associated with a higher risk of divorce.

Roseneil (2000) proposes the same societal transformation through the breakdown or destabilisation of the heterosexual/homosexual binary. She argues that this is due to the breakdown of the hierarchical relationship of the domination of heteronormativity and the decline of the nuclear family. The passing of the UK Marriage Act (2013) extended the legality of marriage to same-sex couples, asserting that it still protects and doesn't diminish religious freedom (Government Equalities Office, 2014).

In an interesting study of Muslim marriage patterns in the UK, Akhtar (2015) recognises a pattern of unregistered Muslim marriages emerging among British Asian Muslims. It has been calculated that approximately 80 per cent of Muslim marriages may in fact not be registered (Duncan Lewis Solicitors Press Release, 2014). The youthful population of Muslims within Britain still hold traditional values concerning matrimony, e.g. many couples do not cohabitate before marriage. Islamic culture states that couples must go through a 'Nikkah' ceremony to have a religiously sanctioned union, and thus be legitimised. These ceremonies have no legal standing within Britain and no legal proceedings of matrimony have followed, states Akhtar (2015). She describes 'Nikkah' as a religious contract which entails the sanctification of marriage, and which establishes the legality of the union in the faith.

The survey conducted by Akhtar (2015) illustrated that Muslim couples feel the need to have their marriage socially accepted by their culture, and placed this at higher importance than being recognised as married by British law, even though this means until they have a legal ceremony, their relationship holds no legal rights. Couples' reasoning included that a legal ceremony had not been convenient for them, and many were happy to continue as they were. Others stated that they would become legally recognised when they have children. Twenty people were involved in answering the survey, which is a modest sample size, but Akhtar unearthed valuable qualitative information in a field of study not well researched.

Feminism, consumption and marriage

Illouz (2003, 2012) has analysed intimacy and relationships across a range of works. Illouz (2012) analyses how in late modernity, capitalism has detrimentally impacted human emotions and intimate relationships by producing individuals who are intrinsically self-centred and career oriented. In *Why Love Hurts*, Illouz (2012) shows how romantic challenges and pain highlight institutional tensions in late modernity. These include the following:

> a mobilisation and refinement of sexual tastes has individualised partner choice, and the monitoring of one's emotions and rational assessments now furnishes the only guideposts for choices disembedded from generalised normative frameworks. Partner choice in the contemporary world thus

becomes at once more rational and more emotional than in previous eras (Nehring, 2013: 1233).

Illouz uses the concept of 'sexual fields' to highlight how sexual tastes in late modernity have become an autonomous and central dimension of partner choice. As Nehring (2013: 1233) comments:

> Sexual fields overlap with, and are structured by competitive marriage markets, through which the commodification of romance proceeds. Here, individuals engage in ongoing competition for the most desirable partners, for the accumulation of partners, and for the display of sexual attractiveness and prowess. . . . Individualised marriage markets, crucially enable individuals to exchange desirable attributes for instance in the form of wealth traded for sexual desirability.

Illouz's conclusions are that there are gender dimensions to this, as it leads to male domination in sexual fields, due to men's ability to linger in them longer and choose from a greater number of partners. As Nehring (2013: 1233) comments: 'Love or the inability to find love thus shapes a dynamic of moral inequalities that divides men and women and the sexually successful from those who fail to achieve'. Nehring is critical of Illouz's position in that it is written at a high level of abstraction and is not embedded in detail of people's lives.

Consumption and the material culture of love: Australian representations of romantic love

As I showed in Chapter 1, the relationship between celebrity, love and marriage is an interesting one, and there is a huge amount of interest in celebrity relationships as well as in the material dimension of those relationships. There is much interest in imitating the material aspects of relationships displayed by celebrities, including jewellery. This can be seen in the public response to Meaghan Markle's jewellery, clothes and accessories. However, this is not a new phenomenon. Boyd (2014) provides an example of this in nineteenth-century novels in Australia as outlined by Teo (2014a):

> Annita Boyd's history of the 'Nellie Stewart bangle', a solid gold bangle given as a symbol of love and commitment to Nellie Stewart, one of Australia's first stage celebrities, by her married lover, George Musgrove in 1885. Boyd's consideration of the material culture of love shows how Stewart's celebrity status ignited a passion for this item of jewellery among young women (Teo, 2014a).

Covering the same period (1880–1930s), Teo (2014b) looks at Australian romance fiction written by women writers, and how they 'conceptualised romantic love, gender relations, marriage and the role of the romantic couple within the nation and the British Empire' (Teo, 2014b). Teo maintains that attitudes towards love and romance changed markedly before and after Australian Federation (1901):

> prior to Australian Federation (1901), short stories about love and romance tended to be more pessimistic about the outcome of romantic love in the colonies. After Federation, however, many of the obstacles to love that had developed in the colonial romance novel persisted, but in the post-Federation romance novel women writers began to imagine that Australian character, culture and environment were sufficient to overcome such obstacles and end happily.
>
> *(Teo, 2014b)*

The influence of the Australian romance novel was enhanced by the Americanisation of Australian culture, as I have shown elsewhere. McAlister and Teo (2017) point out that the influence of publishers such as Mills and Boon was significant in the development of a culture of love and romance. Mills and Boon novels were influenced by their association with Harlequin Press in the US. McAlister and Teo show that Mills and Boon had an arrangement with Harlequin in North America to distribute paperback romances. Perhaps most interesting is the fact Richard and Mary Bonnycastle, the owners of Harlequin, 'liked "pure" or "clean" romances suitable for teenage girls, similar to those produced by North American authors such as Louisa May Alcott, Susan Coolidge, L.M. Montgomery' (McAlister and Teo, 2017: 196).

This view of the content of romance novels was directly conveyed to authors by Alan Boon of Mills and Boon:

> Alan Boon explained to Mills and Boon authors that if they wanted to be distributed in North America by Harlequin, they had to mind their morals, because the Bonnycastles '[d]o not like heroines to be in love with other men, or unhappy married situations, and no touching on differences of colour or novels with "sex" in them, by which he meant intense passion and kissing' (McAleer, 1999, 122–123). The more 'passionate' romances continued to be produced by British authors but they were simply not distributed in North America until the 1970s.
>
> *(McAlister and Teo, 2017: 198)*

In this vein, and in an Australian context, McAlister and Teo (2017: 198) discuss 'Rosa Praed's *The Maid of the River: An Australian Girl's Love Story* (1905), which explores what happens to a young woman who is seduced by her English lover and then abandoned when she becomes pregnant, only to be saved by the laconic and dependable Australian hero at the end.' The novel acts as a romance and a post-colonial moral fable.

One of the features of the Mills and Boon approach was the way in which it addressed the Anglo-centric nature of the romance market in that the novels were about English women immigrating to Australia.

McAlister and Teo show that in more recent Australian novels, there are both discourses of romance and intimacy. They show that this is because:

> romance novels from the 1990s onwards devote more space to articulating the hero's point of view. In Amy Andrews' *Single Dad, Outback Wife* (2007) for example, hero and heroine Andrew and Georgina feel intense passion towards each other: the word 'primitive' is applied regularly to both their sexual attraction and the outback setting. But it is intimacy, established through communication, that allows their romance to be formed . . . Shumway observes that 'Romance offers adventure, intense emotion and the possibility of finding a perfect mate. Intimacy promises deep communication, friendship, and sharing, that will last beyond the passion of new love' (Shumway, 2003: 27).
>
> *(McAlister and Teo, 2017: 213)*

The shift from romance to intimacy has been widely documented (see Brooks, 2017; Shumway, 2003). McAlister and Teo show how this shift is played out in novels in an Australian context by looking at Barbara Hannay's *The Bridesmaid's Best Man* (2008):

> Heroine Sophie and hero Mark already have passion; the novel begins after they have slept together at a wedding in England and Sophie gets pregnant. The subsequent plot revolves around how Sophie comes to Mark's cattle station in the outback so that they can work out parenting arrangements. In the process of doing so, Sophie and Mark become emotionally close as well as physically intimate. The novel thus exemplifies how the discourse of intimacy works alongside the older traditional discourse of romance.
>
> *(McAlister and Teo, 2017: 214)*

While these novels give emphasis to the Western discursive shift from romance to intimacy, there are more specific Australian characteristics conveyed in these novels. In Andrews' *Single Dad, Outback Wife* (2007) and Hannay's the *Bridesmaid's Best Man* (2008), these characteristics include finding 'a place of belonging' which includes 'being in harmony with the land, and in some senses being possessed by it. This is a more post-colonial fantasy: a desire to establish connectedness with the land rather than conquering it or settling it' (McAlister and Teo, 2017: 215).

Migration, marriage and intimacy

In *Distant Love: Personal Life in the Global Age*, Beck and Beck-Gernsheim (2014) discuss the relationship between migration, marriage and intimacy. In a chapter entitled 'Intimate migrations: women marrying for a better life', they look at whether women who migrate to marry are part of a trafficked group of exploited women or women who have made pragmatic decisions to change their lives.

They ask the question of why women migrate for marriage. They show that romantic love is far from the reason which encourages mixed-nationality couples. They point out that sometimes this begins with a marriage agency, personal adverts in a newspaper or a state-organised marriage tour or an internet forum. They argue that many mixed-nationality unions begin not with love, but with the hope of emigrating in order to escape from poverty and the prospects of one's homeland.

They show that a language has grown up around the issues of migrant marriages which are called a variety of things, including 'mail-order brides', 'visa wives' and 'imported husbands'. Beck and Beck-Gernsheim (2014: 78) claim that:

> In politics and the media such marriages are often criminalized; they are suspected of being sham marriages. Feminists frequently see them as part of the worldwide exploitation of women and refer to them as instances of male violence (or more precisely, as part of the pattern in which the dominant Western man exploits the helpless foreign woman).

Beck and Beck-Gernsheim look at the issues of 'commercial matchmaking'. They say that the international matchmaking industry has expanded significantly since the mid-1990s. They say that the forms of broking extend from the internet, newspaper advertisements and organised travel, to sex tourism. They show how South Korea is one such country where mixed-nationality marriages have taken place and, as in Singapore, have been met with hostility.

They also discuss what is called 'chain-migration', which is linked to marriage. Sometimes Beck and Beck-Gernsheim say that pioneer migrants become marriage brokers, and they raise the question of whether women who migrate for marriage are victims or active agents:

> Women who migrate in search of marriage are by no means just feeble, helpless creatures, sold off by men and forced to live abroad. Many have taken this path because they wish to or because they saw no other way to escape a wretched life. Emigrating in search of marriage often results from a definite decision, a conscious act of deliberation . . . staying at home and attempting to build a life there or trying one's luck as a domestic worker. . . . or even a prostitute . . . emigrating in search of a marriage may easily come to seem the more attractive option.
>
> *(Beck and Beck-Gernsheim, 2014: 96)*

'The cultural logic of desire'

Nicole Constable (2005a: 16) has written a great deal on marriage and migration. As she says, 'Women can use marriage mobility to their own advantage and to improve their life prospects'. In *Romance on a Global Stage*, Constable (2003) argues that choice of partner is not the entire picture. In the case of marriage-related migration, the aim is to acquire an entry permit to the First World. However, she does not discount the possibility of romantic motives playing a part.

Constable says that it is part of the 'cultural logic of desire' – that is, people see everything in the same light in that it is everything to do with the West, that is fantasies about Western men. 'Constable (2003: 13) also observed similar trends in the United States with images of "mail-order brides" in the media conveying similar stereotypes of "lotus-blossom" or "dragon-ladies". Constable also shows that the way Asian "brides" are represented in American catalogues is not as "explicitly sexual" but more closely associated with "ideas of love" and "traditional gender roles" (Constable, 2003: 94)' (cited in Brooks and Simpson, 2012: 102).

Elsewhere, Constable (2003) proposes a reconceptualisation of correspondence brides and their male partners in terms of placing the spotlight on the real-life experiences or motivations of those involved. She states it is flawed to always view these women as helpless victims of an oppressive regime. Constable makes the point that women who have worked since childhood in factories and farming to help provide for their families may see staying at home looking after a man in a Western country as being a liberation from drudgery.

The commodification of intimacy and marriage

Elsewhere Constable (2009: 53) provides an interesting overview of how relationships and intimacy have been commodified transnationally. Constable maintains that 'aspects of intimate and personal relations – especially those that are linked to households and domestic units . . . are increasingly and evermore explicitly commodified, seemingly linked to commodities and to commodified global processes . . . or lives that are shaped by market demands that characterize modernity (Beck and Beck-Gernsheim, 1995; Shumway, 2003; Zelizer, 2005).' She goes on to say that intimate relations cover a wide range of areas, including reproductive labour or care work, but also aspects of the entertainment industry, including 'stripping, exotic dancing, hostessing, and other types of sex work' (Constable 2009: 53).

Constable states that there are three areas of intimate relations covered, including cross-border marriages, migrant domestic workers and care workers, and migrant sex workers as they relate directly or indirectly to commodification. She shows how patterns of gendered migration can be found at both ends of the class spectrum: firstly, the high end:

> On one end are elite 'astronaut families' (in which the family members are divided across regions, for example, the male breadwinner may go work in one country, and the wife accompanies the children to another country to facilitate their education) and immigration professionals (Ong, 1999; Piper and Roces, 2003; Shen, 2005, 2008; Shih, 1999; Yeoh et al., 2005), many

of whom have the privilege to choose to remain at home or whose 'home' becomes multiple and flexible.

(Constable, 2009: 55)

But perhaps most work has been focused on the opposite end of the spectrum:

One the other end of the spectrum are growing numbers of documented and undocumented women and men from poorer countries who provide benefits for the more privileged residents of wealthier regions (Ong, 2006) as maids (Anderson, 2000; Adams and Dickey, 2000; Lan, 2006; Parrenas, 2001; Sim and Wee, 2009), sex workers (Augustin, 2007a, 2007b; Cheng, 2007, 2009; Parrenas, 2008), wives (Constable, 2003, 2005a; Oxfeld, 2005; Roces, 2003; Thai, 2005, 2008) or adoptees (Cohen, 2007; Dorow, 2006 . . .) all of whom contribute to what Parrenas has called a 'chain of love'.

(Constable, 2009: 57)

Interestingly, Constable goes on to show that despite being cast at the lower end of the class spectrum, conflicts exist between the class and educational identity of Vietnamese and Chinese wives and their US working-class husbands who seek 'traditional wives' abroad, and women who seek 'modern husbands' (Constable, 2003, 2005b; Thai, 2005, 2008). Much research has been undertaken on these relationships, but there is research on other relational and geographical areas which highlight complexities of class and gender identity.

Maia (2007), in a really fascinating piece of research on Brazilian erotic dancers in New York City, shows how gender and class complexities illuminate the situation. The Brazilian women involved are educated middle-class women who make decisions to act as dancers in night clubs, which presumably pays more than they could earn as maids. In this context, they are more likely to come into contact with larger numbers of working-class and less-educated middle-class US men to whom they could provide intimate labour.

Marriage, consumption and new technologies

The expansion of the Internet has expanded the proliferation of businesses that facilitate arranged marriages. They act as either marital introduction services or to provide services for mail-order brides. Constable (2009: 53) maintains that: 'Internet technology plays a central role in the commodification of intimacy and in shaping new movements and geographic and electronic landscapes of intimacy for individuals (Brennan, 2004; Constable, 2003, 2007)'.

Some marriage brokers promote international marriage partners as though they were commodities or offer services to facilitate the process of meeting and selecting partners from a wider globally defined 'marriage market'

(Constable, 2003, 2005a; Freeman, 2005, 2006; Piper and Roces, 2003; . . . Thai, 2008; Wang and Chang, 2002).

(Constable, 2009: 53, 155)

However, the internet has also influenced how the purchase of sexual services operates, because sex workers now advertise directly online and communicate directly to clients via the internet without the need for middlemen or intermediaries (Augustin, 2007a, 2007b; Bernstein, 2007).

Research has increasingly focused on the authenticity of relationships based on cross-border marriages. Brennan (2004, 2007) approaches Dominican women's relationship with foreign men, some of which result in marriage, as 'performances' in which sex workers feign love to mask the economic exchange and the benefits they receive. Bernstein (2007: 7) argues that '"the girlfriend experience" is increasingly offered by sex workers and often located via the internet is an example of "bounded authenticity" in which not only, eroticism but also an "authentic relationship" (albeit within a bounded frame) is for sale in the marketplace.'

In a study of Filipina entertainers who marry Japanese men, Fairer (2007) focuses on these within the context of 'professions of love'.

> Instead of questioning the authenticity of women's profession of love for their husbands, or treating them as feigned performances. . . . Fairer argues that Filipinas' professions of love serve to counteract the stigma of their work and to define their transnational subjectivities. Love is also associated with the care and understanding offered to women by their husbands.
>
> *(Constable, 2009: 55)*

Constable also makes the point that Fairer's analytical approach to love also has important implications for the debate about migrant agents or victims of trafficking. As Constable (2009: 57) comments:

> Fairer's analysis illustrates unequal global power relations within the context of women's self-definitions. Loving their husbands resonates with their sense of self as moral and modern women and wives. The lure of the Japanese entertainment industry for poor and unemployed Filipinas, the opportunities that such employment offers for intimate socialization with Japanese men, and the shortage of Japanese brides in rural regions of Japan illustrate ways in which capitalist processes promote new opportunities for intimacy and marriage that are influenced by, but not entirely defined by, the entertainment or sex work industry.

Conclusion

Chapter 6 has explored what marriage means and how it operates in contemporary society. The ides of heterosexual marriage as being the universal norm for marriage has long been superseded by a range of different types of marriage globally. In June 2015 in the US (followed by many countries around the world), marriage was recognised as a fundamental right for all, regardless of gender and sexuality. Anthropologists have long shown that several societies functioned either without formal marriage or within different structures. Marriage in the West has changed significantly, and divorce and serial marriage are now the norm. The chapter also reviews different approaches to marriage, as in Muslim marriage patterns in the UK. Transnational marriage, as discussed in the work of Beck and Beck-Gernsheim and more comprehensively by Constable, has provided extensive research on what have been referred to as 'mail-order brides'. Constable challenges the stereotype view of transnational marriages as being exploitative of 'transnational wives'. As shown in this chapter, a range of researchers have provided rich and interesting case studies from the Dominican Republic and the Philippines to highlight a more authentic understanding of women in transnational marriages.

7

ADULTERY, LOVE AND SOCIAL NETWORKING

Introduction

This chapter explores adultery in relationships and in its representation in the media. I examine the significance of adultery in the twenty-first century and assess whether it carries the same moral significance as it has in the past (Leonard, 2010a). I am particularly interested in the role gender plays in adultery and whether the role of women in relationships has fundamentally shifted. Leonard (2010a) examines the relationship between adultery tropes and working women. She also explores how marriage remains important in women's lives, and shows how this can be seen in the multi-billion dollar wedding industry. This chapter explores this in the 'celebrity marriage'. It also looks at how this is represented in film and television, as in *Sex and the City*, and how it combines fashion and celebrity in marriage. Leonard shows how adultery is represented in a number of high-profile Hollywood films and television, including *The Good Wife* and *Unfaithful*. The chapter explores a range of research both classical (Kipnis, 2003) and contemporary (Buunk and Dijkstra, 2006) to assess the historical and contemporary significance of infidelity (Fisher et al., 2010, 2016; Tsapelas et al. 2010).

The chapter also explores the historical context of adultery (Turner, 2002) and shows the significant gendered discourses surrounding adultery (Gregg, 2013). I also examine a more contemporary analysis of 'surveillance' through the concept of 'spouse-busting'. The chapter also considers the theoretical contributions of Kipnis (2003) and Berlant (2011) as well as precarity (Brooks, 2016). The chapter explores the case of *Ashley Madison* in the United States, which casts itself as the website for establishing adulterous relationships.

Adultery tropes and gender

Suzanne Leonard (2010a) uses 'adultery tropes' to unpack class conceptions of working women. She sets out to show that the relationship between female adultery as a cinematic mainstay problematises the figure of the twenty-first-century working woman. As Leonard (2010a: 101) comments, this is an interesting relationship, 'given its imperative to examine the class-specific relation not only to the public sphere of waged labor but also to the seemingly privatized sphere of marriage'.

Drawing on the work of Glass (2004), Leonard highlights the changes that have taken place:

> More women are having affairs than ever before. Today's woman is more sexually experienced and more likely to be working in what used to be male dominated occupations. Many of the affairs begin at work. From 1982 to 1990, 38 percent of unfaithful wives in my clinical practice were involved with someone from work. From 1991 to 2000, the number of women's work affairs increased to 50 percent (Glass, 2004).
>
> *(cited in Leonard, 2010a: 108–109)*

Leonard notes, as shown below, that the sheer number of films being produced on adulterous working women appears to be reaching the level of a 'moral panic', as she comments (Leonard, 2010a: 110):

> Yet the sheer prevalence of discourses on female adultery suggests that at least one strain of cultural anxiety has coalesced around the figure of the adulterous working woman, whose increased visibility has been underscored by the plethora of popular representations that showcase her as a cultural phenomenon.

The backdrop of these films is the fact that heterosexual marriage is still seen as central in women's lives, and this can be seen in the multi-billion dollar wedding industry. There are many examples of this in the media, including films as well as the style channel, which focuses on celebrity weddings such as the Clooney wedding (see http://www.instyle.on/celebrity/george-clooney-and-amal-alamuddins-wedding-photos).

The class and ethnic nature of many of the films developed around this 'cultural anxiety' emerge from the conceptualisation of the 'postfeminist woman' (see Brooks, 1997), as Leonard (2010a: 103) comments:

> the postfeminist working girl appears almost inevitably as a white, upper middle-class woman who is affluent, educated, and urban as exemplified by the paradigmatic figures such as Ally McBeal, Bridget Jones, and *Sex and the City*'s Carrie Bradshaw. Not coincidentally, these women all avoid occupational drudgery, in the sense that they are engaged with jobs that grant them a high degree of cultural currency such as working as a newspaper columnist.

Examples of this kind of film expressing cultural anxiety can be seen in films such as *American Beauty* and *The Good Girl*. Leonard (2010a: 108) says there are at least ten female adultery films distributed in the US, and these include: 'American Beauty (1999), A Walk on the Moon (1999), Lovely and Amazing (2001), The Good Girl (2002), Unfaithful (2002), The Secret Lives of Dentists (2003), Tadpole (2003), Closer (2004), Being Julia (2004).' Leonard (2010a: 110) maintains that the strategy is most apparent in the film American Beauty.

> audiences are encouraged to hate everything the perfectionist Carolyn Burnham (Annette Bening) stands for, namely cold calculating ambition and an utter inability to appreciate her 'good' life or her sweet husband Lester (Kevin Spacey). Clearly meant to represent a career woman whose presentations and greed render her not only a bad wife and mother but also pathetic and humourless, Carolyn's affair with a popular real estate agent (Peter Gallagher) dubbed 'the King' by his fellow agents, is similarly rendered as one propelled not by desire than by ruthless selfishness. In fact, the film suggests Carolyn has sex with Buddy, a vile narcissist, for the basest of motives, namely to further her own career.

Leonard also shows that in addition to the films, there are a number of representations of unfaithful women on television. She notes that in 2004, The Sopranos's Carmela Soprano eventually had an affair after years of tolerating her adulterous husband, and as Leonard notes, her adultery came after her lover suggested she read Madame Bovary. Another example cited the television series Desperate Housewives, which started showing in 2004, and which won Golden Globes in 2005 and 2006 for best comedy. Interestingly, in the somewhat prudish yet sexually exploitative environment in US society, some advertisers withdrew support for the series because of its attack on family values.

Historically, the links between boredom and female adultery have been linked to the trope of unfaithful women driven to what was regarded as scandalous behaviour out of domestic malaise. Leonard (2010a: 112) notes the following:

> Excluded from the sphere of waged labour their boredom was at least partially attributable to the fact that women's duties were for the most part limited to the events of family life. Perhaps the most famous of all bored adulteresses was Flaubert's Madame Bovary, a figure whose extramarital dalliances proved paradigmatic in identifying the idea that adulterous actions might testify to the corrosive effects of marital, familial and domestic boredom. Adultery as a literary and historical construct, might be understood as a mode that articulated the boredom and restlessness that were the inevitable by-product of a social order which mandated women's estrangement from the economies of public labor and productivity. Likewise the adultery genre has been almost unique in its capacity to take seriously the forms of frustration and disappointment characteristic of women whose unhappiness seemed to matter little on the cultural or national stage.

Unfaithful women and the patriarchal order

In a number of films, the female adultery narrative is used, says Leonard, to explore the threat that the unfaithful woman poses to the patriarchal order. An example of this is the film *Unfaithful* (Richard Gere and Diane Ladd). As Leonard says, the film normalises the murder of Diane Ladd's lover by making Gere a highly sympathetic character. The behaviour of the wife is clearly seen as leading her husband to behave in murderous ways. While Gere provides a sympathetic character for audiences, the model of the husband in *The Good Wife* presents a decidedly unsympathetic image.

Elsewhere Leonard (2010b: 3), in discussing the much-acclaimed US television series *The Good Wife*, comments as follows:

> *The Good Wife* is perhaps to be praised for the cynicism with which it approaches the economic and professional calculations that drive these public confessions. Peter's image consultant advises him to seek counsel from a black pastor in order to better connect with a black constituency, and while Peter's religious conversion is initially presented as plausibly genuine, later episodes suggest a more bitter truth. Feigning religiosity, Peter uses his and his family's church-going as means to escape the confines of house arrest and stage a surreptitious meeting with a former colleague.

Leonard draws on the classical theoretical analysis of adultery put forward by Laura Kipnis. Leonard (2010a: 114) comments that Kipnis, in her ground-breaking essay 'Adultery', uses a Marxist theoretical framework and introduces the term 'surplus monogamy' in order to argue that marriage is like work. Kipnis maintains that adultery functions as a protest against marriage, and indirectly against capitalism and the effects of a service economy. She says that marriage operates like a modern workplace.

Adultery – facts and figures

Research shows that historically, there is a relationship between gender and infidelity. This should not surprise anyone. Research in the US indicates that:

> men have a stronger desire to engage in sexual infidelity (Prins et al., 1993) and are more likely to engage in sexual infidelity (Allen and Baucom, 2004; Atkins et al., 2001) . . . have more episodes of infidelity, short or long-term affairs and one night stands (Brand et al., 2007). . . . Husbands are more suspicious of a wife's potential sexual infidelity, as well as more likely to discover a wife's affair (Brand et al., 2007).
>
> *(Tsapelas et al., 2010: 13–14)*

A study by Buunk and Dijkstra (2006) show that American couples indicate that 20–40 per cent of heterosexual married men and 20–25 per cent of heterosexual married women will have an extramarital affair during their life-time. When polled, approximately 2–4 per cent of American men and women had had extramarital sex in the past year.

Tsapelas et al. (2010: 184) maintain that:

> among women, the strength and frequency of affairs are related to the degree of dissatisfaction with the primary relationship, whereas among men the desire to engage in infidelity is less dependent on the state of the primary partnership (Prins et al., 1993). Although women are more distressed about their own infidelity (Van den Eijnden et al., 2000), Allen and Bauman (2006) report that American women are less concerned about hurting their spouse.

Currently, American dating couples report a 70 per cent incidence of infidelity (Allen and Bauman, 2006), and in a recent survey of single American men and women, 60 per cent of men and 53 per cent of women admitted to 'mate poaching', trying to woo an individual away from a committed relationship to begin a relationship with them.

While gender differences in infidelity are now significantly reduced, some aspects of gendered infidelity continues. While women still seem to be more likely to engage in infidelity when they are not satisfied with their primary relationship, men have intercourse with extramarital partners regardless. Same-sex couples show slightly different patterns of infidelity. One study found that gay men were seven times as likely to have sexual encounters outside their primary relationships, compared to heterosexual men (Buss, 2000).

Tsapelas et al. (2010: 176) argue that:

> infidelity was widespread in former decades and in historical and tribal societies. Reports in the 1920s indicate 28 percent of American men and 24 percent of women were adulterous at some point after their wedding. In the late 1940s and early 1950s approximately 33 percent of men and 26 percent of women in an American sample were adulterous (Kinsey et al., 1948, 1953). . . . Data collected in the 1980s suggest that 72 percent of men and 54 percent of women were unfaithful at some point during their marriage. Infidelity was also common among the classical Greeks, Chinese and Hindus (Fisher, 1992).

'Spouse-busting': intimacy, adultery and surveillance technology – Melissa Gregg

Melissa Gregg (2013) has made an interesting connections between intimacy, adultery and surveillance technology. She shows how 'spouse-busting' websites are part of a booming industry that renders marital disloyalty open to both amateur and professional surveillance.

As Gregg (2013: 301) comments:

> The very need for adultery technologies is symptomatic of a period in which some individuals see few options for intimate support – few visions or practices of community- other than the fulfilment to be gained from a dependent partner. As Laura Kipnis argues, the modern relationship is one in which lovers 'must know everything there is to know about one another' (2003: 162). This accords with broader transformations in intimacy encouraging openness and communications between self-directed individuals (Giddens, 1992; Shumway, 2003; Illouz, 2007).

The idea of surveillance and secrecy in relationships is not new, and Turner (2002), in his book *Fashioning Adultery: Gender, Sex and Civility in England, 1660–1740*, notes that monitoring and surveillance strategies were key to proving criminal cases in seventeenth-century England. To partake in intimate relations with another was to partake in a 'criminal conversation'(Turner, 2002: 47).

> Opportunities for men and women to engage in 'wicked', 'illegal' or 'libidinous' conversation increased with wider social changes in the use of public or private space. . . . By the 1700s the transgression of adultery carried specific consequences. For men, the charge conveyed a failure of responsibility and an abuse of authority if the affair was with a servant. For women, adultery generated 'revulsion' among society members who read their actions as upsetting convention through an act of 'domestic rebellion'.
>
> *(Gregg, 2013: 304)*

Turner's study is seen to provide useful historical context to understand the different conditions linked to surveillance. Turner explains that the legal basis of adultery came to be associated with location, namely what Turner calls 'private and suspicious' places (Turner, 2002: 157).

> Early court testimony from neighbours and servants, among others, illustrates that the refusal to open a door, or having a locked as opposed to a 'latched' door, were each taken to indicate 'a crime of secrecy' between conspiring couples (Turner, 2002: 158). Turner's analysis of cuckoldry and adultery trials shows how intimate encounters were increasingly brought to public attention. The affairs of the elite class 'did much to further the opinion that the *beau monde* lived by a code of sexual manners significantly removed from the rest of society (2002: 193)'.
>
> *(Gregg, 2013: 304)*

Gregg notes that the 'crime of secrecy' continues in today's 'spouse-busting' websites which, she argues, list a host of apparently incriminating activities performed by partners. *E-Spy Software* has a set of top signs of a cheating spouse, including: 'a sudden interest in a different type of music'; 'a sudden preoccupation with his or her appearance'; 'an excessive amount of time on the computer when you are asleep' and 'deleting emails'.

New technology can both assist but also expose infidelity, and there are some high-profile cases which show this. Gregg (2013: 305) provides examples of two high-profile cases:

> The case of Anthony Weiner, the Democratic Congressman forced to resign in the wake of a Twitter photo scandal in 2010, shows the loopholes that continue to what 'having sexual relations' with a woman might mean (Berlant and Duggan, 2001). Caught 'Tweeting' a picture of his underpants to a girl he'd met on his work website. . . . The 2012 resignation of CIA Director General David Petraeus following revelations of an ongoing affair is a further example of the political stakes of mediatised infidelity. Leaving aside the curiosity that a CIA head could be so poorly informed about the mechanics of email, the case epitomises the conjunction of morality, transparency and (in this case national) security. As in the Clinton scandal, in high office adultery is taken to be especially significant since it can be placed on a sliding scale of dishonesty that is presumed to end in treason.

Adultery and precarity: *Ashley Madison*

Gregg (2013: 308) maintains that for 'the middle class in an era of what the theorist Bauman (2003) calls "liquid love", a decline in occupational security means the workplace and the domestic unit are equally challenged to provide reliable displays of recognition, identity, and support – all key attributes associated with love and intimacy'.

Berlant (2011: 185) maintains that in a time of precarity, 'love is only slightly less contingent than work'. As Gregg notes, the tagline for the dating website *Ashley Madison* is: 'life is short, have an affair'. As Gregg (2013: 309) comments:

> Anxieties about adultery are always anxieties about security: financial, psychological, ontological. They are worries about losing what little can be counted on; about holding onto an image of someone that we partially know and want to trust but who may not –indeed cannot stay the same.

In 2015, *Ashley Madison* was shown to have created millions of fake profiles. *The Guardian* reported the case over a series of articles (21 August 2015, 31 August 2015, 1 September 2015, 2 September 2015 and 9 September 2015). An article on 31 August 2015 maintains that only 1,492 women have ever checked the site, and that most of the female users are fake 'angel' bots. The details were revealed when 37 million accounts were stolen from the website; despite protestations from the company, the entire company was shown to operate largely as a scam.

An article in *The Guardian* by Sam Thielman, dated 31 August 2015, makes the following comment:

> The latest unattributed statement from the Canadian company came from Annalee Newitz, a writer for Gawker Media websites including Gizmodo, studied the hacked databases for indications that the site's purported 5.5 million female users had arranged trysts using the service. She found that 'only 1,492 women had ever checked the messages' according to the stolen information, though she allowed for the possibility that the data might have been corrupted.

Ashley Madison of course challenged these findings, but the article goes on to say that further research also indicates that *Ashley Madison* created many 'angels' to lift their numbers. In addition, a former employee of the company sued the site in 2013 for repetitive stress injuries to her wrists resulting from the creation of so many fake profiles. In addition, one of the CEOs quit the company because it was revealed he used his work to arrange 'paid assignations' with young women, despite having claimed total fidelity to his wife in an interview with the *New York Daily News*.

In fact, an article in *The Guardian* on 2 September 2015 by Alex Hearn claims that Newitz later retracted the initial claim but reported that 'Ashley Madison created more than 70,000 female bots to send male users millions of fake messages, hoping to create the illusion of a vast play land of available women.' The depth of the scam was shown in *The Guardian* article:

> The bots, Newitz writes, would send users messages designed to entice a response, and then make use of one of the unusual features of Ashley Madison – that men have to pay to read messages, and pay to send them – to encourage the (overwhelmingly male) recipients to buy credits from the site to engage in further conversation.

Newitz writes: 'Ashley Madison aspired to be a global network of people breaking the bonds of monogamy in the name of YOLO. Instead it was mostly a collection of straight men talking to extremely busy bots who bombarded them with messages asking for money.'

Conclusion

Adultery is not a new concept, and what is most interesting about it is not the act of adultery, but its representation in film and social media as an indicator of social change. In this chapter, we have explored how adultery has been conceptualised within a wider social context. As shown by Turner (2002), responses to adultery have moved from viewing it as a criminal case to viewing it as having the potential for exploitation, as in the case of *Ashley Madison*. What is clear is that adultery is a response to the tedium of marriage for many people. Perhaps what is more interesting is the way in which marriage is still retained in the same way in society. As I write this conclusion, the US is addressing the case of a serial adulterous presidency in the form of Donald Trump, whose marital infidelities are now taking on the form of political misdemeanours and campaign violations, in his payments through his then-lawyer Michael Cohen to two women to retain their silence in advance of the 2016 presidential race. The situation involves a porn star, Stormy Daniels, and a Playboy 'Bunny', Karen McDougal, both of whom were paid funds and made to sign NDAs to maintain their silence. While politics rather than morality has been in play here, the interesting issue is whether First Lady Melania Trump will maintain her silence and position as the scandal gets bigger and impeachment looms. Adultery, it has to be said, is never a boring research subject.

8

NEW TECHNOLOGY, INTIMACY AND WORK

Introduction

This chapter looks at new media technology and its impact on intimacy, its impacts on everyday life and how it influences the work-life balance. This chapter looks at Facebook and other social networking platforms to assess what impact this has on everyday life and intimacies. Gregg (2011) considers the relationship between new media technologies and intimacy in relation to people's homes. This chapter also examines the relationship between technology, intimacy and transnational families (Francisco, 2013). In addition, the chapter examines other cultural contexts in relation to intimacy and technology (Hannaford, 2014), including 'intimate surveillance'. Additionally, Parrenas (2005b) raises the issue of 'long distance intimacy' and elsewhere (Parrenas, 2014), examines the constitution of intimacy in the use of communication technology in Filipino transnational families. At a more macro level, the chapter also reflects on the intersection of technology, work and capitalism and its impact on love, intimacy and relationships.

Technology, work and capitalism

It is argued by a number of social theorists that global capitalism relies on the rationalisation of technology to produce a migration industry which commodifies people as products. Francisco (2013: 5) maintains that 'Critical media scholars argue that technology and communication have and continue to be essential to the production process and to the accumulation of capital (Ekman, 2012; Manzerolle and Kjosen, 2012; Pleios, 2012)'. She maintains that historically, technology facilitated the rationalisation of production, circulation and manufactured goods and traffic in people (Appadurai, 1990; Harvey, 1991, see also Brooks,

2014a). It is maintained that an alternative media is challenging the hegemonic globalisation model.

The intersection of technology and migration has always been an aspect of the globalised economy, and the age of neoliberal globalisation has ushered in an increase in international migration. Valerie Francisco (2013), in 'The Internet is Magic: Technology, Intimacy and Transnational Families', argues that the emergence of open-source media, social media, social networks and user-created drives the democratisation of communication through media and technology (Fuchs and Mosco, 2012). Francisco maintains that this process of analysis situates advances in technological innovation in the larger, historical and political economic trajectory of capitalism. Francisco (2013: 8) maintains that: 'Web camera and video conferencing technology not only ushered in more rapid exchange of capital, communication and goods for corporations, but has restructured material and intimate relations for transnational families.' This area will be explored more fully later in the chapter.

Work's Intimacy: Melissa Gregg

The relationship between technology, home, office and relationships is explored by Melissa Gregg (2011) in her book *Work's Intimacy*. Her work focuses on the relationship between the office and homes of workers to provide insights into the personal, family and broader social tensions emerging in today's diverse work environment and its impact on intimacy.

Her research was undertaken in Australia between 2007–2009, and focuses on white-collar workers in the creative, communication and information industries. The focus is on how new media technologies are shaping employees' experience of work at the time. Hamilton (2012: 412) has the following commentary on her research:

> Although Gregg's isn't focused exclusively on those working in the creative or cultural industries her research uncovers the preference for white collar workers to understand their roles as at least nominally creative, and in documenting the highs and lows of flexible, technologically-enabled working lives she reveals the many creative solutions and justifications workers develop to deal with their often complicated work lives.

Gregg shows how telecommunications companies advertised mobility and the collapse of temporal and spatial dimensions as technical possibilities that all workers could take advantage of. Gregg shows that while the new media technologies provide new ways of working, as the office moves into the home office, kitchens, bedrooms, trains and cafes, it also impinges on the personal lives of employees.

In the chapter 'Working from home', Gregg begins to unmask some of the myths around the concept. For example, in the case of working mothers, working from home brings real challenges in that individuals are expected to be reachable at all times. There is a mixed attitude to changes in working life in the last decade. While workers enjoyed the freedoms of working from home and embraced new technologies, they recognised that this involves new pressures. As Hamilton (2012: 413) comments: 'For most participants in Gregg's study the workplace was an unproductive inefficient site not suited to the primary business of working.' Most saw home as a 'consolatory space' accommodating the limitations of the modern workplace.

An important outcome of Gregg's work was the documentation of the rise of Facebook and its changing position in the workplace. Facebook was really developing and still popular at the time of Gregg's study. While Gregg (2011: 105) highlights the benefits of social networking she raises the issue: 'if social media are one of the key means by which employees resist the intrusion of work on their personal lives, what kind of labour politics will be needed to resist management pressures to pilfer friendship networks for business profit'.

Hamilton (2012: 414) makes the point that an important aspect of Gregg's thesis 'is that the popularity of online social networks such as *Facebook* does not merely illustrate the dominance of a culture of commercialised public discourse but in the context of new working lifestyles, the popularity of *Facebook* as a site for intimate contact with friends and family is a response to the fact that demands of work "prevent the likelihood of more significant long term connections beyond the computer screen" (Gregg, 2011: 90)'.

She also considers how bringing work life into the home has an impact on family members:

> As children grow up with work-focused parents, part of their education is to witness the labour regimes that will be necessary to secure their destiny as middle-class professionals.
>
> *(Gregg, 2011: 136)*

In another chapter, 'Home offices and remote parenting', Gregg shows how familial relationships are shaped by work's demands: while men find creating a time and space less challenging, women have to deliberately free themselves from other tasks in order to allow for solitary time for work. Lohmeirer (2012: 1246) comments:

> Melissa Gregg shows clearly that the world of home offices, social networking sites and intensive smart phone use is not free from problems or inconveniences, and most notably it brings new kinds of pressure and expectations -both internally and externally imposed on the individual. Two themes that pervade the monograph are the lack of guidelines from management and the acceptance of individual responsibility by employees.

New technology, intimacy and transmigrant workers

Both Valerie Francisco (2013) and Rachel Parrenas (2005b, 2014) have undertaken very interesting research on how new technologies have influenced relationships of intimacy for transmigrant workers. Francisco examines how new technology in the form of Skype and Facebook 'ushers in a different quality of relationship for transnational families'. She looks at undocumented female transmigrant workers who are unable to leave the US and return to the Philippines because of the legal consequences if they try and gain re-entry. Hence, they have to be away from their families for very long period of time. As Francisco (2013: 3) comments:

> The development of computer technology has changed the lives of migrants and their families left behind. . . . Technology is giving a way for families who are separated over time and space to explore new intimacies through their digital lives, which have many consequences for the ways transnational families are keeping their lives together from afar.

Francisco shows how Filipinas living in New York who have been away for more than ten years have experienced the shift in technology's impact on their family lives. She (2013: 3) outlines the views of a 60-year-old Filipino migrant mother, Carmie, who has lived abroad for over twenty years:

> 'You know, internet is magic. It's magic because before when you write letters it takes months before the receive it. It's the internet that keeps us together. Cam to cam. When I'm not online for two days they worry. So we always talk on Facebook, Yahoo'. Carmie tells of her sustained duty abroad as a migrant mother and worker away from her family through the technological and social media developments in her career as a migrant mother. Her comment also demonstrates the changes in emotional connection that have occurred with the tide of technology. Her statement shows that as the migrant, she is not the sole person in her transnational family that needs and craves communication; it is also demanded and initiated by families left behind in the Philippines.

Developments in social media has meant that communication can take place during the working day for transmigrant women:

> Text messages on mobile phones can be received at any time without migrants having to physically interrupt a workday or conversation to receive information or communication (Lan, 2006). . . Texting is also used for more colloquial and conversational exchanges like jokes, poems riddles and personal messages (Horst, 2006; Lan, 2006). . . Text messages allowed separated families to exchange daily banter that helped them feel closer across long distances.
>
> *(Francisco, 2013: 7)*

Francisco also shows how web cameras and video conferencing have added additional dimensions to communication, and have introduced more elements of surveillance and control into communication. She draws on the work of Derby (2010) in elucidating this situation in Mexican transnational families: 'In her research on Mexican transnational families Joanna Derby found that phone calls alleviated some of the strain of separation, but they also brought up feelings of shame and guilt for migrant parents who were not able to see their young children grow up (Derby, 2010)' (Francisco, 2013: 9).

On the issue of surveillance, Francisco (2013: 9) shows that it acts as a system of control on husbands who commit infidelity:

> Migrant women launching surveillance of husbands and household do so to keep control of the kind of households they would want if they were still in the Philippines. They watch because they would like their children to do their homework and their husbands to prepare meals for the children. . . . And yet for some, surveillance also keeps out a popularity trend in infidelity for Filipino husbands left behind. Infidelity had always existed in Rosie's relationship with her husband even when she lived in the Philippines. As her stay in New York becomes longer and longer, she had been hearing that her husband was up to his old ways again.

Rachel Parrenas (2005b) has a long history of excellent research into emotional labour and transmigrant workers. In what she calls 'long-distance intimacy', she argues that although communication eases the strain in transnational families, it also exacerbates existing gender imbalances in the Filipino family. For migrant mothers, in fact, migration is a no-win situation. Technology opens up the opportunity for migrant women to do the care-work from afar, and they are able to instruct women at home to assume their roles. Parrenas's (2005b) article 'Long Distance Intimacy: Class, Gender and Intergenerational Relations between Mothers and Children in Filipino Transnational Families' shows how Parrenas links new technology and intimacy:

> I do this by identifying and examining the transnational communication methods Filipino migrant families use to develop intimacy, in other words familiarity, across borders. In my analysis, I address how political economy and gender shape the dynamics of transnational communication. By showing how economic conditions and gender shape the dynamics of transnational communication. By showing how economic conditions and gender shape transnational family communication
>
> *(Parrenas, 2005b)*

In a later article titled 'Intimate Labour of Transnational Communication', Parrenas (2014) examines the constitution of intimacy in the use of communication technology in Filipino transnational families. It argues that intimacy in transnational families often has a different ontology than in nuclear families: the former, she argues, is frequently defined by routine, while the latter is frequently characterised by instantaneity. This article criticises the failure of literature on transnational families to recognise this difference, which has resulted in the expectation of transnational mothers to maintain intimate relations that embody nuclear family characteristics. Mothers are expected to be 'here and there', 'absent-present' and always available at a distance. Parrenas maintains that such an expectation overlooks the ways in which migrant women have reconstituted mothering in transnational families, and retains the ideology of female domesticity.

Perhaps more fundamentally, it is the constitution of gender in intimate labour practices of migrant parents in transnational families which is at the core of the article. Parrenas shows that it is the ideology of female domesticity and the idea that women must nurture, retain proximity with and be involved in the day-to-day lives of their children. Other researchers have shown the same thing in Filipino, Honduran, Mexican and Salvadoran transnational families (Parrenas, 2005b; Shamalzbauer, 2005; Derby, 2006; Abrego, 2009). As she shows in relation to the work of Derby and Abrego:

> Distinguishing the distinct parental expectations of transnational mothers and fathers, Derby (2006) for instance noticed the greater emotional involvement expected of mothers in Mexican transnational families. As she states, 'mothers' relationships with their children in Mexico are highly dependent on demonstrating emotional intimacy from a distance, whereas fathers' relationship lie in their economic success as migrant workers (Derby, 2006: 34). Likewise, Abrego (2009) observed that Salvadoran transnational mothers firmly maintained their caregiving from afar via their selfless commitment to their children's wellbeing. Men, she noted, did not.
>
> *(Parrenas, 2014: 427)*

Parrenas shows that gendered expectations for women in transnational families is impossible to achieve particularly in relation to 'instantaneity' due to structural constraints that hinder access to communication technologies. While some (Madianou and Miller, 2012) argue that the new media provides a solution, however imperfect, to the cultural contradictions of mothering and migration, others maintain (Derby, 2010: 204) that the gender dynamics of Mexican transnational families show that: 'Children evaluate mothers more harshly for having left them. Migrant mothers bear the moral burdens of transnational parenting'. In other words, they are the ones who have to establish an 'absent presence'.

Parrenas (2014: 429) shows that not all migrants are able:

> to perform the intimate labour of forging interpersonal ties with their children in the Philippines via instantaneous communications. As noted by Francisco (2013) in her study of domestic workers in New York, some do. But in my research, including the interviews I conducted in June and July 2013 with 48 Filipino transnational mothers in the United Arab Emirates

transnational families . . . those that can instantaneously communicate are still rare.

Similarly, the research undertaken by Liebelt (2011) in Israel observes that: '[o]n Sundays, the usual weekly day off for Filipino live-ins, Filipinos queue to talk to family members and friends in the Philippines in front of the public phone booths'. Likewise, planned telephone calls had also been the common mode of communication in the Mexican transnational families observed by Derby (2010) in Mexico and the US.

There are significant structural challenges for transnational communities in terms of communication. Internet is not widely accessible for migrant workers, and telephone costs are prohibitive. Text messaging is less expensive, but requires contractual obligations. As Francisco (2013: 434) notes:

> As Derby (2010: 62) describes of Mexican transnational families 'many do not have a home phone in Mexico and call their children at a neighbour's home or at a local business'. Likewise my interviewees in Dubai were almost all from remote areas of the Philippines, where as I described . . . many do not have access to running water at home.

In addition, as previously mentioned, mothers are more expected than fathers to maintain communication regularly. Francisco notes that a migrant father is expected to fulfil his duties by being a breadwinner. In contrast, migrant mothers had to maintain an 'absent presence' in order to be a good mother.

Many migrant workers are also found in Singapore, and the work of Platt et al. (2004) explores the use of information technology among Indonesian Foreign Domestic Workers (FDWs) working and living in Singapore. Drawing upon a survey of 201 domestic workers, and a follow-up of 38 in-depth interviews, the research points to changes in the technological landscape in Singapore which have facilitated access to new technology. This includes cheaper mobile devices and increased access to free internet, either through work or in the public sphere. The argue that these changes have reframed the experiences of foreign domestic workers in three key areas: firstly, it has allowed foreign domestic workers to more easily straddle the transnational divide so they can increase communication with new friends and family. Secondly, access to Wi-Fi at the employer's home is based on trust, as it involves access to passwords. Thirdly, they argue that such access to information technology provides a form of social empowerment by allowing global contact. However, as shown above, there are significant caveats to these points.

Surveillance and control in the use of technology in transnational marriages

In their assessment of the value of technology in communication between husband and wife in Senegalese transnational marriages, wives of migrant husbands complain about the kind of control that husbands can exert through the use of communication technology. Hannaford (2015: 43) comments that: 'Instead of enabling "emotional closeness" the virtual presence of their absent husbands frequently represents a spectre of suspicion, control and surveillance'.

Hannaford uses case studies of Senegalese wives to show how women in transnational marriages can come to represent and even dread the virtual presence of their husbands. As Hannaford (2015: 44) shows:

> While some migrant husbands return periodically, most cannot afford frequent trips to Senegal to visit their wives and family. They conduct their marriages over the phone, through email, skype, wire transfers, and instant messages. . . . In these marriages, communication technology is not a simple instrument with which to bring wives closer to absent husbands. Rather the constant threat of surveillance that such technology enables puts a great deal of stress on these transnational couples.

Hannaford (2015: 45) uses the example of Ndeye to highlight the issues:

> Ndeye and her husband were married for 11 years before he moved abroad to work in Ohio. She says that they have a happy marriage but that the distance is difficult for them both. Ndeye's husband has opted to leave his wife and children in Senegal rather than to apply for family reunification from the USA government and to bring them over to live with him, which is a relatively common choice for Senegalese migrants (Gasparetti and Hannaford, 2009). He prefers to send money to them in Senegal where they can live more cheaply and where his children can benefit from the language and the values he holds dear.

When migrant husbands visit home, the situation is pleasant and fun; however, in contrast, when they are overseas, their attitude is jealous and possessive, as Hannaford (2015: 45) shows:

> Ndeye's husband has a work schedule and salary that allows him to come home to visit her and their three children for three months every two years. When he comes home on vacation, Ndeye says they live like newlyweds, going to the beach or out dancing in nightclubs with live music. Ndeye puts extra money into cooking his favourite dishes and the mood in the house is celebratory. When he is gone however Ndeye says that her husband becomes very jealous. He calls her randomly throughout the day on the house's landline phone to ensure she is always at home.

The lengths husbands go to check on the movement of their wives is extraordinary, as Ndeye's situation attests:

> She confessed that when her husband called her mobile phone when she is with neighbours, she lies and tells him she is at home. Recently, her husband

started to test her by asking to speak to one of the children. Ndeye's neighbours have seen her sprint up the street towards home screaming her son's name, hoping to convince her husband that she is merely calling to the back of the house. Ndeye and many of the wives of migrants find themselves effectively tethered to their homes by the telephone cord.

(Hannaford, 2015: 46)

Another woman in Hannaford's study, Adama, also suffers from a controlling husband who denies her permission to visit relations and friends:

She says the surveillance does not even stop when she travels to the rural region of her origin for a baptism or some other ceremony. Her husband will call someone else at the event and ask him or her to pass him to his wife. When she is at home and needs to make a trip downtown or to visit a friend or relative she has to ask his permission. The other day she wanted to visit her aunt who is very ill, but her husband reasoned that other relatives who lived closer could take care of her and denied permission. If he does agree to let her go somewhere he will call her mobile at some point during the day to tell her it is time to go home. Then call the house every half an hour until she gets home to know exactly the hour she returned. 'Every half an hour' she repeated with emphasis. 'You can't lie. If you tell him you arrived at a certain time he will tell you what minute he called and no-one answered' (Hannaford, 2015: 46).

Perhaps the most extreme example is Rama, whose husband worked in France and she managed a restaurant in Senegal. However, he began to resent her work at the restaurant.

She explained that because he had met and fallen in love with her at the restaurant he hated the idea that other men might meet her the same way. He wanted to make sure that she was dressed modestly and wearing no make-up to enhance her beauty. Indeed Rama was incredibly beautiful. Even among Senegalese women, who are generally poised and attractive, Rama was especially striking and graceful. Finally the restaurant began to fail . . . and eventually closed altogether to her husband's satisfaction.

(Hannaford, 2015: 53)

Her husband's behaviour is both obsessive and controlling, and he never gives her money because he fears she will use it for taxis to leave the house and meet men. He also does not allow her to have a mobile phone and calls her on the landline to make sure she is at home. For most women in the West, this would be an intolerable situation and they might well leave, but Rama's world is so confined that she has few options. Hannaford (2015: 53) comments as follows:

In the middle of telling her story, Rama opened a drawer beside her bed and pulled out a gold ring. She explained that her husband takes off his wedding ring before he goes back to France and leaves it in Senegal. 'Do you find that normal?' she asked me. Unlike her husband, Rama has no sense of what goes on in her husband's life in France. He gives her few details and gets angry when she asks. Although they phone, Skype or text one another daily, she has no clear picture of his life abroad and like many migrant husbands he shares very little information with her.

Beyond the usual available technology in the process of transnational communication there are additional dimensions available which act to control those left behind. As Hannaford (2015: 55) shows:

> Mobile phones and landlines are not the only tools that control these women. Voice-over-internet technology, instant messenger software and the increased ease of electronic money transfers further enhances an absent husband's ability to circumscribe his wife's independence. Rama's webcam allows her husband to see her wardrobe and assess its appropriateness. MSN lets him monitor any potential socializing she might do and limits any attempts she might make to use the internet to escape her housebound solitude. Rama's husband can easily use Western Union to send money to her parents to maintain their goodwill towards him, to his friends who then run errands for her and she believes perhaps to her neighbours so that they can be his eyes and ears while he is away.

Hannaford (2015: 56) concludes that many of us might wonder why these women remain in marriages which cause them such distress. Although the divorce rate is rising steadily (Dial, 2008) and its stigma slowly decreasing, divorce is still an unattractive option for most women, particularly when, as in Rama's case, they feel they will not have the support of their families once the marriage ends. Furthermore, divorce in Senegal is complicated for women. Men must agree to grant a divorce for it to take place. If they do not, they must appear in court, which is not easy when they reside abroad. No legal recourse is available to women without a civil marriage (such as those who only married in a mosque). They can only cajole their husbands into agreeing to a divorce. Many women I interviewed had endured long, painful divorces from their migrant husbands, which took years to complete. During the process, the women neither received remittances from their estranged spouses nor, of course, were able to remarry.

Conclusion

This chapter looks at the relationship between new media technology, work, love and intimacy. The chapter reviews the work of Melissa Gregg, and considers whether new technology which has allowed people to work from home has provided greater flexibility or intrusiveness into people's lives and relationships. Gregg points out that the unstructured nature of home work and the lack of responsibility of employees poses problems for workers and puts pressure on individuals to be continually available. The chapter also looks at a fascinating area of new technology: transmigrant workers and relationships. The work of Valerie Francisco provides interesting insights on Filipina transmigrant workers in New York, and shows how new technology has impacted their relationships with their families and their husbands, as well as how new technology can provide a form of surveillance of unfaithful husbands. Parrenas also provides interesting information on how new technology has provided additional pressures for transmigrant mothers, who are, in Parrenas's terms, expected to be 'absent-present' and always available. Finally, Hannaford's research on Senegalese transnational marriages, where the husbands are the transmigrant worker, provides fascinating insight into how husbands use different forms of new technology to control the movement of their wives who are in Senegal; this includes everything from making sure they stay at home to controlling what they wear. Hannaford shows how difficult it is for women to divorce their husbands in Senegal.

9

THE COMMERCIALISATION OF INTIMACY

Introduction

This chapter examines the relationship between romance and consumption (Illouz, 1997), and also explores conceptualisations of postmodern love, which Illouz sees as captured in Candace Bushnell's (1996) *Sex and the City*. The chapter thus focuses on the relationship between *Sex and the City*, consumer culture and postfeminism as an example of one aspect of the commercialisation of intimacy and considers what a postfeminist woman-centred drama amounts to. The chapter also explores the relationship between postmodern romance and the emergence of the love affair; it also considers the changing nature of love affairs in gendered terms and the representation of the adulterous woman in films. The chapter looks finally at the sexualisation of modern relationships, and examines feminist theorists' analysis of the commercialisation of intimacy. Illouz's (2014) *Hard Core Romance: Fifty Shades of Grey* examines a range of controversial and popular literature which focuses on intimacy.

How intimacy is structured by its encounters with late capitalism

In *Consuming the Romantic Utopia: Love and the Cultural Contradictions of Capitalism*, Illouz (1997) shows that definitions of emotions, intimacy and love underwent a significant change within and outside marriage in the move to modernity.

Much has been written about changes in the way in which love and intimacy changed from the eighteenth century onwards in the United Kingdom (Giddens, 1992; Brooks, 2017) and the impact on the self. However, Illouz focuses on how these changes took place in America. She shows that while the Victorian family of the nineteenth century was characterised by an emphasis on piety and devotion to children, intimacy was absent from Victorian marriages.

Illouz argues that the collapse of the Victorian marriage in the twentieth century is seen as a result of pressures including secularisation, consumption and the emergence of democratic expectations of relationships. In her book *Consuming the Romantic Utopia: Love and the Cultural Contradictions of Capitalism*, Illouz (1997: 25) explores '[w]hat role the cultural motive of romance play in the construction of mass markets of consumption and, vice versa, how did romantic practices incorporate economic practices of the market?'

Illouz (1997: 27) shows that romantic love became prominent in America as early as the eighteenth century. She notes the following:

> Americans, more so than citizens of other Western, industrializing societies tended to base their marital choices on emotional considerations rather than social calculations, a tendency that accelerated toward the end of the nineteenth century. The choice of mate was left to the individuals involved because love was deemed to be of paramount importance for conjugal bliss. . . . [The] relaxation of Victorian sexual mores among the middle classes; dating and petting now appeared as accepted forms of sexual exploration prior to marriage.

She shows that the increasing levels of women's education and their entry into the workforce explain, at least partially, weakened adherence to Victorian ideology: 'The change in sexual mores and the increasing push toward the equalization of men's and women's status in the public sphere in turn affected the ways in which they spent their leisure time: they began to engage in the same leisure pursuits and became full and equal members of what historian Kathy Peiss has called a "heterosocial world"' (Illouz, 1997: 27).

The commercialisation of intimacy

Peiss's work is an interesting reflection on the relationship between consumption and beauty in the nineteenth and early twentieth century, and the impact on women as professionals and in relation to changes in the social mores around love and intimacy. Peiss (2011: 4–5) shows that the beauty industry was built by women:

> In the early stages of developing the cosmetics industry from the 1890s to the 1920s, women formulated and organized 'beauty culture' to a remarkable extent. The very notion of femininity, emphasizing women's innate taste for beauty, opened opportunities for women in this business, even as it restricted them elsewhere. And women seized their chances, becoming entrepreneurs, inventors, manufacturers, distributors, and promoters. . . . The beauty trade they developed did not depend upon advertising as its impetus. Rather, it capitalized on patterns of women's social life – their old customs of visiting conversation, and religious observance, as well as their new presence in shops, clubs and theatres.

In addition, Peiss's work also has implications about the class and ethnic background of women within the beauty industry. As Peiss (2011: 5) shows:

> Strikingly, many of the most successful entrepreneurs were immigrant, working-class, or black women. Coming from poor, socially marginal backgrounds, they played a surprisingly central role in redefining mainstreams ideals of beauty and femininity in the twentieth century. Focusing new attention on the face and figure, they made the pursuit of beauty visible and respectable. In many ways, they set the stage for Madison Avenue, whose narrowly drawn images of flawless beauty bombard us today. . . . Elizabeth Arden was a Canadian immigrant and 'working girl' who remade herself into a symbol of haute femininity; she carved a 'class market' for cosmetics by catering to the social prestige and power of wealthy and upwardly mobile white women. In contrast such black entrepreneurs as Madam J. Walker and Annie Turnbo Malone promoted a form of beauty culture entwined in the everyday lives of poor American women.

Illouz focuses on the fact that new technologies restructured the cultural landscape of early twentieth-century America, with inventions such as telephone, typewriter, high-speed printing press, phonograph, radio, photography, and motion pictures expanding the general public's access to mass culture through newspapers, magazines, popular songs and film.

Alongside this, women's role in society was changing in the public realm and was the subject of criticism. As Peiss (2011: 6) comments:

> Their changing status as workers, citizens, consumers, and pleasure seekers was acknowledged cosmetically. During the nineteenth century, the 'public woman' was a painted prostitute; by its end, women from all walks of life were going public. Women crowded onto trolleys, promenaded the streets, frequented the theatres, and shopped in the new palaces of consumption. . . . A new 'marriage market' substituted dating for courtship, and the dance hall for the front porch; a new sense of sexual freedom emerged.

Illouz shows how alongside changes in the cultural environment were transformations in the meaning of love which emerged from a number of factors, including:

> the extrication of love from religion, that is, the secularization of the discourse of love; the increasing prominence of the theme of love in mass culture, especially in film and advertising; the glorification of the theme of love as a supreme value and the equation of love with happiness; the association of love and consumption, more specifically, the romanticization of commodities; the inclusion of 'intensity' and 'fun' in the new definitions of romance, marriage, and domesticity.
>
> *(Illouz, 1997: 28)*

Love, romance and intimacy were also framed by the way the movie industry represented relationships. Illouz (1997: 32) claimed that the images drawn from the consumer marketplace developed the intersection of love and consumption:

> The Hollywood system established in the interwar period [was significant in this regard]. Many works of most popular film-maker of this period, Cecil B. De Mille, for example, focused on the 'dos and dont's of a successful marriage' . . . the message conveyed by his movies was that men expected women to be beautiful . . . and that women not only expected marriage to be entertaining but waited for their husbands to provide the entertainment.

Illouz makes the point that during the first quarter of the twentieth century, the theme of romance became increasingly associated with consumption. As Illouz (1997: 37) shows:

> Thus the theme of romance was used for the promotion of a wide array of products, a process I called the 'romanticization of commodities'. In these images, the 'aura' of romance impregnates two levels of consumption. The first is the consumption of the product being advertised, what I called *candid consumption*. But romance is also associated with another type of consumption, never explicitly referred to: the activity in which the couple is engaged which often consists in the consumption of leisure. I term this last category of consumption-equated with romance but never made explicit – *oblique consumption*.

Refashioning love from pre-modern to postmodern love

Illouz is concerned with the relationship between romance and consumption, as outlined above, in her early work, and she developed this relationship more fully in her later work. Illouz (1998, 2010) discusses the refashioning of love from pre-modern to postmodern love, and she shows that modern love and relationships are more emancipated and egalitarian than their pre-modern counterparts.

The shift in emphasis is described as 'disenchantment', and Illouz describes pre-modern society as 'enchanted love', opening up the self 'to a quasi-religious sense of transcendence'. Illouz (2010: 22) sees postmodern love as being captured in Candace Bushnell's (1996) *Sex and the City*. Bushnell expresses in the following extract 'a thoroughly self-conscious supremely ironic, and disenchanted approach to love':

> When was the last time you heard someone say, 'I love you!' When was the last time you saw two people gazing into each other's eyes without thinking, yeah right? When was the last time you heard someone announce, 'I am truly madly in love,' without thinking, 'just wait until Monday morning?'
>
> *(Bushnell, cited in Illouz, 2010: 22)*

Illouz describes this as a 'chick lit' genre geared towards women and focused on the problems inherent in relationships. Within this genre, love has become the privileged site for the trope of irony. As Illouz (2010: 22) comments: 'The rationalization of love is at the heart of the new ironic structure of romantic feeling, which marks the move from an "enchanted" to a disenchanted cultural definition of love'.

Sex and the City, consumer culture and postfeminism

In an interesting article on the ground-breaking nature of *Sex and the City,* Jane Arthurs (2003), in '*Sex and the City* and consumer culture: remediating postfeminist drama', describes *Sex and the City* as a postfeminist, woman-centred drama and addressed an audience which was 'white, heterosexual, relatively youthful and affluent'. She maintains that the objective was to maximise audiences by creating a drama that could appeal to both women and men.

The distinction between work and private lives is somewhat blurred in the television series (as opposed to the films) of *Sex and the City.* It is not clear whether the relationship between these areas was ever completely thought through. Arthurs (2003: 84) establishes the relationship as follows:

> In *Sex and the City,* the world of work largely disappears from view as a distinct space and set of hierarchical relations, although the women's autonomy from men is underwritten by their economic independence. For three of the four women who make up the main characters in the series, work is collapsed into the private sphere and becomes another form of self-expression alongside consumption, thereby side-stepping the postfeminist problematic. Carrie's sex life and those of her friends act research for her weekly newspaper column, which she writes from home. Samantha works in public relations, a job where her physical attractions and personal charm are intrinsic to her success. Charlotte manages an art gallery in a manner that suggests it is more of a hobby. . . . Only Miranda feels the contradiction between her private life and her career success as a lawyer.

Arthurs notes that the 'traditional romance narrative is still there, but as a residual sensibility, a slightly old-fashioned version of femininity that doesn't work in practice' (Arthurs, 2003: 85). For example, in the case of Charlotte, whose views of marriage represent a largely eighteenth-century perspective, her belief in romance and in saving herself for her husband is confounded by the fact that she finds her husband impotent and his arousal requires porn magazines – a lovely ironic touch. As Arthurs (2003: 87) comments: 'Feminist evaluations of *Sex and the City* have conflated it with other examples of postfeminist culture in which comedy and satire have replaced any serious ethical commitment to challenging the power relations of patriarchy, a challenge they argue is undermined by complicit critiques'.

The self-mocking attitude of *Sex and the City* is also expressed in another scene where Carrie (Sarah Jessica Parker) gathers her friends together for the launch of a new publicity campaign which promotes her weekly column called 'Sex and the City'. As Arthurs (2003: 86) notes:

> They wait on the sidewalk for a bus to pass by carrying the poster for her Brand on its side. They are in a mood of excited anticipation, marred only by the regret that Mr Big, the new man in her life had failed to show up to share this proud moment. There is the poster with Carrie's body stretched in languorous pose along the full length of the bus, under the strapline 'Carrie Bradshaw knows good sex.'

However, there is also an obscene graffiti drawn on top of the advert. As Arthurs (2003: 87) comments: 'The idealised image of bourgeois perfection in the image of Carrie on her billboard is momentarily satirised by the obscene graffiti.' In fact, it is a feature of much of *Sex and the City* that it combines an overt commitment to consumer culture with a resistance to cultural hegemony. As Arthurs (2003: 88) comments:

> *Sex and the City* self-consciously explores the instability of feminine identity as a postfeminist postmodern consumer culture . . . *Sex and the City* exemplifies these features of the commodity. Its stylistic features contribute to the cultural hegemony of the incorporated resistance of the bourgeois bohemians. Its cultures of femininity provide an alternative to heterosexual dependence but its recurring promise of a shameless utopia of fulfilled desire always ends in disappointment for the cycle of consumption to begin again.

Arthurs has it about right when she shows that 'the transgressions of bourgeois sexual decorum' by the four women can be categorised as 'unruly' and a challenge to patriarchal structures, but as she says: 'their adherence to the sleek control of the commodified body makes this compatible with capitalism' (Arthurs, 2003: 92).

Because Arthurs understands the power dimension implicit in *Sex and the City*, she reflects on Carrie's frustrations as being power based. As Arthurs says, Carrie explains her powerlessness in relation to 'Big', and she [Arthurs] (2003: 95) 'offers a poignant critique of the masquerade as a strategy of female empowerment' as she repeats Carrie's comments:

> 'I think I'm in love with him, and I'm terrified in case he thinks I'm not perfect . . . you should see what I'm like around him – it's like I wear little outfits. I'm not like me. Sexy Carrie, Casual Carrie. Sometimes I catch myself actually posing . . . it's exhausting!'
>
> *(Episode 11, 'The Drought')*

The apparent incongruity between Carrie's independence and well-networked life, and her continual sense of having to accommodate 'Big', her boyfriend, is never fully resolved. *Sex and the City* is very much a product of its time; there is little social diversity, and it would be unthinkable now to have a series coming out of the US with four white women whose discourse is restricted to consumption and men. There is no ethnic dimension in the series, although a brief concession is made in one of the films to include an African-American woman. Gay relationships provide a brief interlude in one of the films, as one of Carrie's male friends gay, but there is little or no discussion about lesbianism other than a passing comment from Samantha. Despite its limitations, *Sex and the City* has provided an important discussion point for feminists, and it raised a large number of issues regarding sexual subjectivities and women's agency.

Bushnell's *Sex and the City* provided an ideal example for Illouz to develop her relationship between love and intimacy, consumption and postmodernism. Illouz has a cultural model of love, showing how cultural forces have refashioned love and have contributed to its disenchantment. She identifies science and internet technology as part of this change, as well as psychological knowledge and what Illouz (2010: 27) identifies as 'technologies of choice' that emerge with the growth of the market of choice:

> the premodern actor looking for a mate seems a simpleton in comparison with today's actors, who from adolescence to adulthood develop an elaborate set of criteria for the selection of a mate. . . . Psychology, internet technology, and the logic of the capitalist market applied to mate selection have contributed to create a self-conscious, manipulable personality, who uses an incredibly refined and wide number of criteria, presumably conducive to greater compatibility.

Illouz undertakes a study of online dating in 2006, and she shows that: 'the majority of respondents reported that their tastes changed in the course of their search and that they aspired to "more accomplished" people than they did at the beginning of the search.' In other words, Illouz maintains that people use rational strategies to achieve their romantic aspirations. As Illouz (2010: 28) comments:

> Like no other technology, the internet has radicalized the notion of the self as a 'chooser' and the idea that the romantic encounter should be the result of the best possible choice. That is, the virtual encounter has become hyper-cognized, the result of a rational method of gathering information to select a mate. It has literally become organized as a market, in which one can compare 'values' attached to people, and opt for 'the best bargain.' The internet places each person searching for another in an open market of open competition with others, thus radicalized the notion that one can and should improve one's romantic condition and the partners are eminently interchangeable.

The romanticisation of commodities and the commodification of romance

Illouz's analysis of the relationship between romance and consumption in the twentieth century is analysed by Turner (1998), who describes her work as a 'sociology of romance' through 'the romanticisation of commodities'. As shown earlier, the romanticisation of commodities describes the way in which romance became popularised in the US through the growth of advertising and the development of a mass market for romance.

Illouz sees romance as part of the growth of leisure and consumption. The growth of the film industry and the role of romance in film is part of this. Turner (1998: 116) states that 'The use of close-ups in film and photography and the employment of movie stars to advertise commodities created a new cosmology in which icons represented the new lifestyle'. The growth of dating as a social phenomenon against the backdrop of the growth of youth culture, including 'dining out', became part of how dating and courtship were defined.

Commodification of love is part of what Illouz claims is part of the American Dream. Part of this is the introduction of the leisure sphere into romance. Illouz (1997) shows how this is focused on the 'romantic holiday'. She shows how this is a key example of the commodification of romantic love. Illouz sees this as part of a liminal experience, separate from home and work, and intensifying intimacy. Typical adverts like the Sandals advert remain hugely potent as a romantic experience which encourages love and romance. The following are typical of the way Sandals is advertised:

> www.sandals.co.uk/website indicated/
> A Paradise Created. For Two People in love.
> Enjoy all inclusive luxury holidays at Sandals' Caribbean resorts and see for yourself why we offer the most romantic getaways with more quality inclusions than any other luxury beach resort. With gorgeous white sand beaches, savory dining experiences, opulent accommodations, unlimited refreshing beverages, exciting water sports and more. It's wonder why so many choose Sandals as their tropical escape.

Turner (1998), in a critique of Illouz's analysis, maintains that while Illouz acknowledges the cultural contradictions in contemporary love, she fails to acknowledge the significance of adultery as a universal feature of contemporary marriage.

Postmodern romance: the love affair

In *The Lost Innocence of Love: Romance as a Postmodern Condition*, Illouz (1998) argues that postmodern romance has seen the collapse of overarching, life-long romantic narratives, which it has compressed into a briefer and repeatable form of the affair. As Illouz (1998: 176) claims:

> In contradistinction to the teleological, absolute and single-minded Romantic narrative of 'grand amour,' the affair is a cultural form that attempts to immobilize and repeat compulsively, the primordial experience of 'novelty'. Moreover, the affair is undergirded by a consumerist approach to the choice of a mate. During the Victorian era, people chose from a very narrow pool of available partners and often compelled to marry their first suitor.

As Illouz goes on to say this does not mean that the affair has the anarchic and 'disorganized' character often postulated by postmodern theorists. Quite the contrary: most people's sexual partners belong to a pool of people who are quite similar to them, and this despite the fact that barriers of locality, race and religion have broken down. Illouz (1998: 184) shows this also suggests that the 'postmodern affair' is quite different from the indiscriminate and unbridled search for pleasure and the motive of male domination that characterise Don Juan's or Casanova's sexual quests.

Illouz recognises that the rise of the affair is directly related to the transformation of the role of sex in relationships since the Second World War. She shows how sex as a central aspect of pleasure in relationships was legitimised by feminist and gay groups. The principle on which the affair is based is choice and a market framework within which to choose.

Illouz (1998) shows that the experience of 'waiting' which was the pattern of Victorian women's lives is now eliminated. This is replaced by periods of sporadic romantic intensity which Illouz says have the character of postmodern fragmentation. Illouz (1998: 177) notes: 'Affairs then, are self-contained narrative episodes disconnected from one another in the flow of experience, resulting in a fragmenting of experience of love into separate emotional units'.

Illouz makes it clear that sex has always been a feature of relationships outside marriage, but she claims that the character of these contemporary affairs is distinctively postmodern for a number of reasons. She says firstly that, like many other factors in postmodernism, they 'institutionalize liminality'(Illouz, 1998). In other words, the affairs are located away from home and work, removed from marriage, family and domestic responsibilities.

Secondly, Illouz claims that the character of sexual pleasure is more about 'liberation' than previous embodiments of sexual pleasure, and she says that such affairs are likely to be enjoyed by both sexes. Thirdly, Illouz says the affair is not a deliberate statement of transgression, and they do not set out to challenge normative or moral imperatives. The final point made by Illouz is that the underpinning of the postmodern affair is how identity is being redefined based on choices and consumer rationality. The affair is, of course, linked to adultery and whereas traditionally adultery was seen as being initiated by men, increasingly women are seen as initiating adulterous affairs.

'Architectures of choice' and the emergence of marriage markets

The purpose of Illouz's (2012) analysis of love and intimacy in late modernity is to shift the emphasis away from Freudian and psychological explanations of the breakdown of marriage and the failure of love, to a set of social and cultural tensions and contradictions which have impacted the self and identities. Illouz's (2012: 4) feminist view of love and marriage is one which sees power as the basis of relationships; '[the claim] made by feminists is that a struggle for power lies at the core of love and sexuality and that men have had and continue to have the upper hand in that struggle because there is a convergence between economic and sexual power'.

Illouz makes some interesting observations of the relationship between love and patriarchy. She maintains that when patriarchy was powerful, love was less significant in relationships. But she says that with the decline of patriarchy, there has been an increase in 'the cultural prominence of love'. One of the most transformative aspects of love, Illouz claims, is the capacity for choice, and she describes this as 'architectures of choice'.

In explaining this, Illouz claims that it is conditions within which choices are made which has transformed relationships. The second factor was a change in the way potential partners are assessed, to include both psychological and sexual dimensions, with the emphasis being on the latter. The third factor is the increasing potency of 'sexual fields' – in other words, the importance of sexuality in determining relationships, and in acting as a significant factor in choices being made in the marriage market.

For Illouz, the central feature of the modern era is this intersection of consumer culture and desire, which she sees as at the centre of subjectivity. She sees sexuality and desire as interchangeable. The burgeoning of a huge market for consumer products linked to the emergence of the market around beauty and glamour reinforced the emphasis on sex and desire. The range of choices available to individuals is not one-dimensional, and can be a mixture of economic, emotional and sexual traits.

Illouz's (2014) *Hard-core Romance: Fifty Shades of Grey, Best Sellers and Society*, can be seen as a 'self-help' book, entering the mass market and capitalising on 'giving publicity to what has become a crisis of sexual privacy' (Illouz, 2014: 36). Illouz claims that through the examination of contemporary romantic or 'chick lit' literature, it can be seen that not only do ideas of love, romance, emotions and intimacy enter the public sphere of mass consumption and popular culture, but also offer guidance to women.

Fifty Shades of Grey does produce different responses from some feminists, who argue that it 'represents a shift of cultural ideas and gender roles back towards the traditional role of submissive female and the dominant male'(Purcell, 2013). Women's literature is critiqued as representing the sexual objectification of women and discarding female liberation and equality. However, Illouz (2014: 77) maintains that:

> self-help is not only a market segment, it is a whole new modality to culture; that is, it constitutes a new way by which the individual connects to society. Because modernity entails a very large amount of uncertainty about self-worth, and the norms and morality that should guide relationships, self-help becomes one of the main pathways for the shaping of self-hood.

Crispin (2014) considers an overview of Illouz's (2014) book *Hard Core Romance: Fifty Shades of Grey, Best-Sellers and Society*, and examines what makes a book a bestseller and what a bestseller reveals about a culture. She shows that how an idea catches on in a society means that it cannot be a radical one. It must, says Crispin, 'constellate thoughts and experiences that people are already having and yet are perhaps not yet able to articulate'. She looks at what these ideas are. Illouz refers to women's mass culture as a self-help culture, and her examples include Oprah fiction and women-focused magazines and talk shows.

Illouz identifies this realm where women are meant to locate themselves in relation to understanding 'their relationships, their bodies and their psyches – is where *50 Shades* got its start' (Crispin, 2014). Crispin maintains that Illouz understands that 'self-help seems like a kind of masculinized competitiveness, in a different and more anxious mode'. Across many books written by Illouz, including *Why Love Hurts; A Sociological Explanation* to *Saving the Modern Soul: Therapy, Emotions and the Culture of Self-Help* to *Oprah Winfrey and the Glamour of Misery*, she highlights that this is how one derails feminist movements, by turning societal problems into individual failures. As Crispin (2014) comments: 'In this mode, the source of inequity turns into psychological inadequacy . . . it's your personal chemical imbalance that keeps you depressed, not a very real and unhealthy shift in the way we manage our families, our communities, our cities.'

Illouz differentiates herself as a feminist from other commentators in that she focuses her analysis of *Fifty Shades of Grey* on the power dynamic that she sees as the entire basis of the novel. As Crispin (2014) comments:

> Christian has all the power, in the forms of sexual experience, wealth, age and his high-level job, and Anastasia uses her powerlessness to paradoxically gain access to his power. Her virginity, her sexual inexperience, her youth, her formlessness, her emotions, are used as weapons, and in the end she, in a way, dominates him.

Illouz frames the issue of the relationship between intimacy and consumption as revealing a conflict between the anxiety of freedom versus the desire for intimacy and the search for meaning in a consumerist marketplace.

Conclusion

The final chapter of the book provides a detailed analysis of the work of Eva Illouz, who has developed the relationship between romance and consumption across a range of socio-cultural contexts. Her analysis is best captured in the concept of 'the commercialization of intimacy'. The chapter reviews a range of concepts and analyses developed by Illouz across her work. Illouz's work reflects on the intersection of popular cultural discourses through film and television, the beauty industry, literature and social media. As such, her work spans a range of cultural discourses, and traces love and romance from a pre-modern to a post-modern era. Turner (1998) accurately captures her work as being 'a sociology of romance' through 'the romanticization of commodities'. Her insights make the mundane aspects of romance into something coherent and socially authentic, whether she is discussing the 'romantic holiday', the love affair or the marriage market. Illouz never fails to theorise her work within a feminist framework, but one which has the capacity to understand social change in a broader cultural context.

CONCLUSION

Throughout this book we have explored the contemporary nature of love and intimacy in relationships. We have drawn on sociological, feminist, anthropological, historical and cultural explanations for the changing nature of love and intimacy.

One of the interesting dynamics of the changing nature of love and intimacy is how the expression of love and intimacy and relationships more generally are reflected historically. Bogle (2008: 12) frames the 'evolution' of love, intimacy and relationships as moving from 'the calling era' through the 'dating era' and now to the contemporary 'hooking up era'. As Bogle (2008: 12) comments:

> According to social historian Beth Bailey, for the first decade of the twentieth century 'respectable' young men would 'call' on respectable young women at the home. The object of the call was to spend time with the woman of interest as well as her family, especially her mother. . . . Young women and their mothers controlled the practice of calling. That is, they and only they could invite a young man to come to their home for a calling visit. Such a visit typically consisted of spending time in the woman's parlor with her and her family.

There are a number of interesting dimensions to this, including the ability to control relationships and the power wielded by women in this arrangement. There are also class dimensions to this system of calling:

> As entrenched as the calling system was among middle class circles, this script did not work for the lower or working classes. Most members of the lower class lacked the facilities to entertain young men in their homes.

Thus lower-class youth ultimately stopped trying to aspire to the middle-
and-upper-class system of calling.

(Bogle, 2008: 13)

As regards dating Bogle says that the term 'date' can be traced to the late nineteenth century, when it was first used as a slang term by members of the working
class. Bogle (2008: 13) said that it referred to occasions when a man obtained
sexual favours from lower-class women. She goes on to say:

> Dating emerged next among rebellious upper-class youth who began
> going out, away from the watchful eyes of parents. . . . Dating was not a
> matter of upper-class rebellion only, but also grew out of changes in soci-
> ety. Women at this time in history were becoming increasingly part of the
> public sphere, with growing numbers attending college, taking jobs, and
> in general becoming more a part of the public world that was still largely
> considered the province of men.

Bogle notes that dating spread throughout US culture from the first decade of
the twentieth century until about the mid-1920s, and became a 'universal cus-
tom in America'.

The situation was changing socially to advantage women, particularly in education, but this did not advantage them in dating:

> Although women started attending college in greater numbers during this
> period, there was still a six-to-one male-to-female ratio on campus at the
> time. . . . For instance, dating was almost exclusively carried on by frater-
> nity men. Freshmen were not allowed (by tradition) to date co-eds, and
> women from outside were 'imported' for some of the bigger occasions on
> campus.

Despite the fact that 'dating' started as a concept within the working class, it was
soon appropriated by the middle class, and both men and women wanted to date
someone who had 'rank'. As Bogle (2008: 15) states:

> Competition for dates was fierce, and the 'Class A' men on the rating
> scale wanted to be sure only to be seen with 'Class A' women, and vice
> versa. . . . To remain in high standing, women consistently had to date
> Class A men only. Women also avoided drinking in groups or frequenting
> the beer parlors. Women's prestige on campus would decline once they
> were no longer a fresh face on campus, due to indiscretions, or if they were
> too readily available for dates.

The difference between 'calling' and 'dating' was interesting in terms of gender
power. On the issue of 'calling' women and mothers played a dominant role in

decision making, however with 'dating' men led the process and it was a man's right to ask a woman out. Dating also moved into the public realm and away from the watchful eyes of parents.

'Hooking-up' is the latest phase in the evolution of love and intimacy. This shift may have escaped many of us who still proceeded with a 'dating model' but as Bogle (2008: 20) comments this trend started from the 1960s:

> In the mid-1960s, changes in the way young people were getting together had begun to occur. This shift away from traditional dating was particularly apparent on college campuses. College students began socializing in groups, rather than pair dating, and 'partying' with large numbers of friends and classmates. Parties represented more than just a social outing; they became the setting for potential sexual encounters.

Of course, the context of the establishment of relationships must be seen in a broader social and cultural context. These included areas we are very familiar with: 'The advent and increased availability of the birth control pill coupled with a liberalization of attitude towards sexuality led to changes in what was socially acceptable to do sexually . . . intercourse became thought of as a sign of intimacy with physical pleasure rather than merely a means of reproduction' (Bogle, 2008: 21). In addition, there were cultural changes, including the women's movement and changes in the nature of sexuality.

Bogle also notes that the cultural changes underway during the 1960s can be seen to be partially reflected in wider demographic changes which ran parallel to the cultural changes. Bogle (2008: 22) outlines the significance of these as follows:

> a number of demographic trends are relevant to understanding why hooking up emerged and formal dating declined on college campuses. First, there has been an increase in the median age for first marriage is approximately 25 for women and 27 for men. This contrasts with 1960, when the median age at first marriage was approximately 20 for women and 23 for men. Thus the number of people getting married during their college years or immediately after has sharply declined in the past 40 years.

The implications of these changes is that there is less pressure on students to find a partner during their college years; thus, relationships can be more casual than previously. Young people are sexually active by age seventeen, so relationships may be more casual, including sexual relationships.

I set out in this book, firstly, to understand contemporary thinking about love, intimacy and emotions and how they impact relationships in society, as well as to understand the diversity and range of partnerships and relationships in society and to show that there is no one single understanding of relationships. Secondly, I aimed to provide an understanding of different perspectives of love and intimacy from sociological, feminist, filmic, cultural studies and gender studies perspectives, and explore how these perspectives help us understand and resonate with lived experiences of love, intimacy and emotion in contemporary society. Thirdly, I set out to provide a critique and analysis of these perspectives and the key concepts that emerge from these models for understanding the social world. Finally, I aimed to outline the role of social networking and new technology in providing new ways of understanding how relationships are formed.

I have aimed throughout the chapters of the book to develop these ideas, and I hope you have enjoyed reading this book as much as I enjoyed writing it.

BIBLIOGRAPHY

9 1/2 (1986). 'Dir Adrian Lyne'. www.youtube.com/watch

Abbott, S. and Jermyn, D. (eds.) (2009) *Falling in Love Again: Romantic Comedy in Contemporary Cinema*. New York: IB Tauris and Co, Ltd.

Abrego, L. (2009) 'Economic Well-Being in Salvadoran Transnational Families: How Gender Affects Remittance Practices', *Journal of Marriage and Family*, 71: 1070–1085.

Adams, K.M. and Dickey, S. (eds.) (2000) *Home and Hegemony: Domestic Service and Identity Politics in South and South-East Asia*. Ann Arbor: University of Michigan Press.

Akhtar, R.C. (2015) Unregistered Muslim Marriages: An Emerging Culture of Celebrating and Conceding Rights. In Miles, J., Mody, P., and Probert, R. (eds.) *Marriage Rites and Rights*. Oregon: Hart Publishers, 168–192.

Alberti, J. (2013) '"I Love You Man": Bromances, the Construction of Masculinity, and the Continuing Evolution of the Romantic Comedy', *Quarterly Review of Film and Video*, 30: 159–172.

Allen, H. (2007) *"Sensitive and real macho all abtto same time": Young heterosexual men and romance men and masculine*, 10(2): 137–152, oct1, 2007.

Allen, E.S. and Baucom, D.H. (2004) 'Adult Attachment and Patterns of Extradyadic Involvement', *Family Processes*, 43: 467–488.

Allen, E.S. and Bauman, D.H. (2006) 'Dating, Marital and Hypothetical Extradyadic Involvements: How Do they Compare?, *The Journal of Sex Research*, 43(4): 307–317.

Alleyne, R. (2008) 'Romantic Comedies Make Us "Unrealistic about Relationships", Claim Scientists', *The Telegraph*, 15 December. www.telegraph.co.uk/culture/film/Romantic-comedies-make-us-unrealistic-about-relationships-claim-scientists

Anderson, B. (2000) *Doing the Dirty Work: The Global Politics of Domestic Labour*. New York: Zed Books.

Ansari, A. (2015a) 'Everything You Thought You Knew about L-O-V-E Is Wrong', *Time Magazine*, 4 June. http://time.com/aziz-ansari-modern-romance/

Ansari, A. (2015b) *Modern Romance*. New York: Penguin Books.

Appadurai, A. (1990) 'Difference and Disjunction in the Global Cultural Economy', *Theory, Culture and Society*, 7(2): 295–310.

Arnett, J.J. (2004) *Emerging Adulthood: The Winding Road from the Late Teens through the Twenties*. Oxford: Oxford University Press.

Arthurs, J. (2003) 'Sex and the City and Consumer Culture: Remediating Postfeminist Drama', *Feminist Media Studies*, 3(1): 83–98.

Atkins, D.C., Baucon, D.H., and Jacobson, N.S. (2001) 'Understanding Infidelity: Correlates in a National Random Sample', *Journal of Family Psychology*, 15: 735–749.

Augustin, L.M. (2007a) *Sex at the Margins: Migration, Labour Markets and the Rescue Industry*. London: Zed Books.

Augustin, L.M. (ed.) (2007b) 'Special Issue on Cultural Studies of Commercial Sex', *Sexualities*, 10(4): 473–488.

Austen, J. (1811) (1992) *Sense and Sensibility*. New York: Penguin Books and Random House.

Austen, J. (1813) (2000) *Pride and Prejudice*. New York: Modern Library Classics.

Austen, J. (1814) (2003) *Mansfield Park*. Northants: Penguin Classics.

Austen, J. (1815) (2005) *Emma*. New York: Simon and Schuster.

Australian Federation (1901). 'National Library of Australia'. www.nla.gov.au

Barbalet, J. (2002) 'Introduction: Why Emotions are Critical', *The Sociological Review*, 50(82): 1–9.

Bauman, Z. (1995) *Life in Fragments: Essays in Postmodern Morality*. Oxford: Blackwell.

Bauman, Z. (2003) *Liquid Love: On the Frailty of Human Bonds*. Cambridge: Polity Press.

Beck, U. and Beck-Gernsheim, E. (1995) *The Normal Chaos of Love*. Cambridge: Cambridge University Press.

Beck, U. and Beck-Gernsheim, E. (2002) *Individualization*. London: Sage.

Beck, U. and Beck-Gernsheim, E. (2014) Globalization of Love and Intimacy: The Rise of World Families. In *Distant Love, Personal Life in a Global Age*. Cambridge: Polity Press, 4–19.

Bell, D. and Binnie, J. (2000) *The Sexual Citizen: Queer Politics and Beyond*. London: Wiley-Blackwell.

Berlant, L. (2001) Love, a Queer Feeling. In Dean, T. and Lane, C. (eds.) *Homosexuality and Psychoanalysis*. Chicago: University of Chicago Press.

Berlant, L. (2011) *Cruel Optimism*. Durham, NC: Duke University Press.

Berlant, L. and Duggan, L. (eds.) (2001) *Our Monica, Ourselves: The Clinton Affair and the National Interest*. New York: New York University Press.

Berlant, L. and Warner, M. (2000) Sex in Public. In Berlant, L. (ed.) *Intimacy*. Chicago: University of Chicago Press.

Bernstein, E. (2007) Temporarily Yours: The Sale and Purchase of Bounded Authenticity. In Padilla, M. (ed.) *Caribbean Pleasure Industry: Tourism, Sexuality and AIDS in the Dominican Republic*. Chicago: University of Chicago Press.

Bjornberg, U. and Kollind, A.K. (2005) *Individualism and Families*. Abingdon: Routledge.

Bogle, K. (2008) *Hooking Up: Sex, Dating and Relationships on Campus*. New York: New York University Press.

Boucher, L. and Reynolds, R. (2017) Same-Sex Love in Late Modern Australia: On the Political Straight and Narrow. In Teo, H.-M. (ed.) *The Popular Culture of Romantic Love in Australia*. Melbourne: Australian Scholarly Publishing, 341–367.

Boyd, D. (2014). *Its Complicated: The Social Lives of Networked Teens*. New Haven: Yale University Press.

Brand, R.J., Markey, C.M., Mills, A., and Hodges, S.D. (2007) 'Sex Differences in Self-Reported, Infidelity and Its Correlates', *Sex Roles*, 57: 101–109.

Brannen, J. and Nielson, A. (2005) 'Individualization, Choice and Structure: A Discussion of Current Trends n Sociological Analysis', *Sociological Review*, 53: 412–428.

Brennan, D. (2004) *What's Love Got to Do with It? Transnational Desire and Sex Tourism in the Dominican Republic*. Durham, NC: Duke University Press.

Brennan, D. (2007). 'The ABC of Child Care Politics', *Australian Journal of Social Issues*, 42(2): 213–225.

Brooks, A. (1997) *Postfeminism: Feminism, Cultural Theory and Cultural Forms*. London and New York: Routledge.

Brooks, A. (2006) *Gendered Work in Asian Cities: The New Economy and Changing Labour Markets*. London: Ashgate.

Brooks, A. (2010) *Social Theory in Contemporary Asia*. London and New York: Routledge.

Brooks, A. (2014a) "The Affective Turn" in the Social Sciences and the Gendered Nature of Emotions: Theorizing Emotions in the Social Sciences from 1800 to the Present. In Lemmings, D. and Brooks, A. (eds.) *Emotions and Social Change: Historical and Sociological Perspectives*. London and New York: Routledge.

Brooks, A. (2014b) *Popular Culture: Global Intercultural Perspectives*. Basingstoke: Palgrave Macmillan.

Brooks, A. (2016) 'Gender, Precarity and Sexuality in Neoliberal Society: The Influence of Lauren Berlant', Conference Paper Presented at International Sociological Association Conference, University of Vienna, 10–14 July, Vienna.

Brooks, A. (2017) *Genealogies of Emotions, Intimacy and Desire: Theories of Changes in Emotional Regimes from Medieval Society and Late Modernity*. London and New York: Routledge.

Brooks, A. (forthcoming 2020) *The Sociology of Emotions: Feminist, Cultural and Sociological Perspectives*. Bristol: Bristol University Press.

Brooks, A. and Devasayaham, T. (2013) *Gender, Emotions and Labour Markets: Asian and Western Perspectives*. London and New York: Routledge.

Brooks, A. and Simpson, R. (2012) *Emotions in Transmigration*. Basingstoke: Palgrave Macmillan.

Bushnell, C. (1996) *Sex and the City*. London: Abacus.

Buss, D.M. (2000) *The Dangerous Passion: Why Jealousy is as Necessary as Love and Sex*. New York: The Free Press.

Butler, J. (1990) *Gender Trouble: Feminism and the Subversion of Identity*. London and New York: Routledge.

Butler, J. (2002) 'Is Kinship Always Already Heterosexual?', *Differences*, 13(1): 14–44.

Buunk, A.P. and Dijkstra, P. (2006) Temptation and Threat: Extradyadic Relations and Jealousy. In Vangelsti, A.L. and Perlman, D. (eds.) *The Cambridge Handbook of Personal Relationships*. New York: Cambridge University Press, 533–555.

Cacioppo, J.T., Cacioppo, S., Gonzaga, G.C., Ogburn, E.L., and Tyler, J.V. (2011a) 'Marital Satisfaction and Breaks Ups Differ across On-line and Off-line Meeting Venues', *Proceedings of the National Academy of Sciences*, 110(47): 18814–18819.

Cacioppo, J.T., Hawkley, L.C., Norman, G.J., and Berntson, G.G. (2011b) 'Social Isolation', *Annals of the New York Academy of Sciences* 12(31): 17–22.

Cancian, F. (1987) *Love in America: Gender and Self Development*. New York: Cambridge University Press.

Castells, M. (1997) *The Power of Identity*. Oxford: Blackwell.

Charles, N., Davies, C.A., and Harris, C. (2008) *Families in Transition: Social Change, Family Formation and Kin Relationships*. Bristol: Policy Press.

Cheng, S. (2007) Romanticising the Club: Love Dynamics Between Filipina Entertainers and GIs in U.S. Military Camp Towns in South Korea. In Padilla, M. (ed.) *Caribbean Pleasure Industry: Tourism, Sexuality and AIDS in the Dominican Republic*. Chicago: University of Chicago Press.

Cheng, S. (2009) *Dreams of Flight: Sex, Love and Labour of Migrant Entertainers*. Ithaca, NY: Cornell University Press.

Cocks, H.G. (2009) *Classified: The Secret History of the Personal Column*. London: Random House.

Cohen, F. (2007) Tracing the Red Thread: An Ethnography of Chinese-US Transnational Adoptions. Unpublished PhD thesis: University of Pittsburgh.

Cohen, P. (2013) 'Marriage Is Declining Globally: Can You Say That?', *Family Inequality*, 12 June.

Cole, J. and Thomas, L.M. (2009) *Love in Africa*. Chicago: University of Chicago Press.

Connell, W.F. (1957) *Education: An Introductory Survey*. Canberra: Australian Council for Educational Research.

Connell, W.F., Francis, E.P., and Skillbeck, E. (1957) *Growing Up in an Australian City: A Study of Adolescents in Sydney*. Melbourne: Australia Council for Educational Research.

Conor, L. (2013) 'The "Lubra" Type in Australian Imaginings of the Aboriginal Woman from 1836–1973', *Gender and History*, 25(2): 230–251.

Constable, N. (2003) *Romance on a Global Stage*. Berkeley, CA: University of California Press.

Constable, N. (ed.) (2005a) *Cross-Border Marriages: Gender and Mobility in Transnational Asia*. Philadelphia: University of Philadelphia Press.

Constable, N. (2005b) A Tale of Two Marriages: International Matchmaking and Gendered Mobility. In Constable, N. (ed.) *Cross-Border Marriages: Gender and Mobility in Transnational Asia*. Philadelphia: University of Philadelphia Press, 166–186.

Constable, N. (2006) 'Brides, Maids and Prostitutes: Reflections on the Study of "Trafficked" Women', *Portal: Journal of Multidisciplinary International Studies*, 3(2). http://epress.lib.uts.au/ojs/index.php/portal

Constable N. (2007) Love at First Sight?: Visual Images and Virtual Encounters with Bodies. In M. B. Padilla (ed.) *Love and Globalization: Transformations of Intimacy in the Contemporary World*. Nashville: Vanderbilt University Press.

Constable, N. (2009) 'The Commodification of Intimacy: Marriage, Sex and Reproductive Labor', *Annual Review of Anthropology*, 38: 49–64.

Coontz, S. (2006) *Marriage, a History: How Love Conquered Marriage*. New York: Penguin Books.

Coontz, S. (2015) 'Revolution in Intimate Life and Relationships', *Journal of Family Theory and Review*, 7: 5–12. stephaniecoontz.com

Coontz, S. (2016) 'Marriage Is Not What It Seems', In *The Way We Never Were*. New York: Basic Books.

Creed, B. (1993) *The Monstrous-Feminine: Film, Feminism, Psychoanalysis*. London and New York: Routledge.

Crispin, J. (2014) 'Feminism and the "50 Shades" Hangover', *Los Angeles Review of Books*, 15 July. www.lareviewofbooks.org

Croll, E. (2006) 'The International Contract in the Changing Asian Family', *Oxford Development Studies*, 34: 478–491.

Crow, G. (2002) *Social Solidarities*. Buckingham: Open University Press.

De Beauvoir, S. (1972) *The Second Sex*. Harmondsworth: Penguin Books.

De Boise, S. and Hearn, J. (2017) 'Are Men Getting More Emotional? Critical Sociological Perspectives on Men, Masculinities and Emotions', *The Sociological Review*, 65(4): 779–796.

Dempsey, K. (2002) 'Who Gets the Best Deal from Marriage: Women or Men?', *Journal of Sociology*, 38(2): 91–110.

Derby, J. (2006) 'Honor and Virtue: Mexican Parenting in the Transnational Context', *Gender and Society*, 20: 32–60.

Derby, J. (2010) *Divided by Borders: Mexican Migrants and Their Children*. Berkeley, CA: University of California Press.

Dial, F.B. (2008) Manage of divorce a darkar: itiner aves feminism paris: katara

Dorow, S.K. (2006) *Transnational Adoption: A Cultural Economy of Race, Gender and Kinship*. New York: New York University Press.

Duncan, S. and Smith, D. (2006) 'Individualization Versus the Geography of "New" Families', *Twenty-First Century Society*, 1: 167–189.

Durkheim, E. (1897) *Suicide: A Study in Sociology*. New York: Free Press.

Ekman, M. (2012) 'Understanding Accumulation: The Relevance of Marx's Theory of Primitive Accumulation in Media and Communication Studies', *Triple C*, 10(2): 156–170.

Elias, N. (2000) *The Civilizing Process: Sociogenetic and Psychogenetic Investigations*, trans. Edmund Jephcott, ed. Eric Dunning, et al. Oxford: Blackwell.

Elley, S. (2015) 'Confronting a Culture of "Laddishness" and "Riskiness" in Higher Education'. http://discoversociety.org/2015/05/05/confronting-a-culture-of-laddishness-and-riskiness-inhigher-education

Ellinghaus, K. (2002) 'Margins of Acceptability: Class, Education and Interracial Marriage in Australia and North America', *Frontiers*, 23(3): 55–75.

Ellinghaus, K. (2003) 'Absorbing the "Aboriginal Problem": Controlling Interracial Marriage in Australia in the Late 19th and Early 20th Centuries', *Aboriginal History*, 27: 183–207.

Elliott, S. and Umberson, D. (2008) 'The Performance of Desire: Gender and Sexual Negotiation in Long-Term Marriages', *Journal of Marriage and Family*, 70: 391–406.

Enker, D. (1994) Australia and the Australians. In Murray, S. (ed.) *Australian Cinema*. St Leonards: Allen and Unwin, 210–225.

Erickson, R.J. (2005) 'Why Emotion Work Matters: Sex, Gender and the Division of Household Labour', *Journal of Marriage and Family*, 67: 337–351.

Evans, M. (1998) '"Falling in Love with Love is Falling for Make Believe": Ideologies of Romance in Post-Enlightenment Culture', *Theory, Culture, Society*, 15(3–4).

Evans, M. (2003) *Love: An Unromantic Discussion*. Cambridge: Polity Press.

Fairer, L. (2007) 'Filipina Migrants in Rural Japan and Their Professions of Love', *American Ethnology*, 34: 148–162.

Farrow, R. (2017) 'From Aggressive Overtures to Sexual Assault: Harvey Weinstein's Accusers Tell their Stories', *The New Yorker*, 10 and 23 October. http://newyorker.com/news/news-desk/from-aggressive-overtures-to-sexual-assault-harvey-weinsteins-accusers-tell-their-stories

Farrow, R. (2018) 'Les Moonves and CBS Face Allegations of Sexual Misconduct', *The New Yorker*, 6 and 13 August. www.newyorker.com/magazine/2018/08/06/les-moonves-and-cbs-face-allegations-of-sexual-misconduct

Farrugia, R. (2013) *Facebook and Relationships: A Study of How Social Media Use Is Affecting Long Term Relationships*. New York: Rochester Institute of Technology.

Fatal Attraction (1987, Director Adrian Lyne). www.youtube.com/watch

Ferguson, A. (1989) *Blood at the Root: Motherhood, Sexuality and Male Dominance*. London: Pandora.

Ferguson, A. (1991) *Sexual Democracy: Women, Oppression and Revolution*. Boulder: Westview Press.

Ferguson, A. (2012) Romantic Couple Love, the Affective Economy, and a Socialist-Feminist Vision. In Schmitt, R. and Anton, R. (eds.) *Taking Socialism Seriously*. New York: Lexington Books.

Ferguson, A. and Jonasdottir, A.G. (2014) Introduction. In Jonasdottir, A.G. and Ferguson, A. (eds.) *Love: A Question for Feminism in the Twenty First Century*. New York and London: Routledge.

Finkel, E., Eastwick, P.W., Karney, B.R., Reis, H.T., and Sprecher, S. (2012) 'Online Dating; A Critical Analysis from the Perspective of Psychological Science', *Psychological Science in the Public Interest*, 13, 1: 3–66.

Firestone, S. (1970) *The Dialectic of Sex: The Case for a Feminist Revolution*. New York: Morrow.

Firestone, S. (1972) *The Dialectic of Sex*. London: Paladin.

Fisher, H. (1992) *Anatomy of Love: A Natural History of Monogamy, Adultery and Divorce*. New York: Simon and Schuster.

Fisher, H. (2009) 'How to Make Romance Last', *O the Oprah Magazine*, December. www.oprah.com

Fisher, H., Brown, L.L., Aron, A., Strong, G., and Mashek, D. (2010) 'Reward, Addiction and Emotion Regulation Systems associated with Rejection in Love', *Journal of Neurophysiology*, 104(1): 51–60.

Fisher, A. and Manstead, A. (2000) The Relation Between Gender and Emotions in Different Cultures. In Fisher, A. (ed.) *Gender and Emotion: Social Psychological Perspectives*. Cambridge: Cambridge University Press.

Fisher, H., Xiaomeng Xu., Aron, A., and Brown, L. (2016) 'Intense, Passionate, Romantic Love: A Natural Addiction? How the Fields that Investigate Romance and Substance Abuse Can Inform Each Other', *Frontiers of Psychology*. https://doi.org/10.3389/fp.syg.2016.000687

Flaubert, G. (1856) (2013) *Madame Bovary*. New York: Penguin Books and Random House.

Flexner, E. (1972) *Mary Wollstonecraft. A Biography*. New York: McCann and Geoghegan.

Forrest, S. (2010) 'Young Men in Love: The (Re)making of Heterosexual Masculinities through Serious Relationships', *Sexual and Relationship Therapy*, 25: 206–218.

Foucault, M. (1978) *The History of Sexuality Vol 1*. London: Penguin Books.

Francisco, V. (2013) '"The Internet is Magic": Technology, Intimacy and Transnational Families', *Critical Sociology*, 41(1): 173–190.

Freeman, C. (2005) Marrying Up and Marrying Down: The Paradoxes of Marital Mobility for Chosonjok Brides in South Korea. In Constable, N. (ed.) *Cross-Border Marriages: Gender and Mobility in Transnational Asia*. Philadelphia: University of Philadelphia Press.

Freeman, C. (2006) Forging Kinship Across Borders: Paradoxes of Gender, Kinship and Nation between China and South Korea. PhD thesis: University of Virginia.

Freestone, R. and Veale, S. (2004). 'Sydney, 1901: Federation, National Identity and the Arches of Commemoration', *National Identities*, 6(3): 215–231.

Friedman, M. (2003) *Autonomy, Gender, Politics*. New York and Oxford: Oxford University Press.

Fuchs, C. and Mosco, V. (2012) 'Introduction: Marx Is Back: The Importance of Marxist Theory and Research for Critical Communication Studies Today', *Triple C*, (2): 127–140.

Galasinski, D. (2004) *Men and the Language of Emotions*. London: Palgrave Macmillan.

Gasparetti F. and Hannaford F. (2009) Genitorialite a distance: reciprocate & migration senegalese. model migrante, 1: 111–131.

Giddens, A. (1991) *Modernity and Self-Identity: Self and Society in the Late Modern Age*. Cambridge: Polity Press.

Giddens, A. (1992) *The Transformation of Intimacy: Sexuality, Love and Eroticism in Modern Societies*. Cambridge: Polity Press.

Giddens, A. and Sutton, P. (2013) *Sociology 7th edition*. Cambridge: Polity Press.

Gill, R. (2007) 'Postfeminist Media Culture: Elements of a Sensibility', *European Journal of Cultural Studies*, 10(2): 147–166.

Gill, R. and Herdieckerhoff, E. (2006) 'Rewriting the Romance: New Feminism in Chick Lit', *Feminist Media Studies*, 6(4): 487–504.

Glass, S.P. (2004) *Not 'Just Friends': Protect Your Relationship from Infidelity and Heal the Trauma of Betrayal*. New York: Free Press.

Goldberg, A.E. (2013) '"Doing" and "Undoing" Gender: The Meaning and Division of Housework in Same-Sex Couples', *Journal of Family Theory and Review*, 5: 85–104.

Government Equalities Office (2014) *Marriage (Same Sex Couples) Act: A Factsheet*. Gov U.K. www.gov.uk

Greer, G. (1970) *The Female Eunuch*. New York: McGraw-Hill.

Gregg, M. (2011) *Work's Intimacy*. Cambridge: Polity Press.

Gregg, M. (2013) 'Spouse-Busting Intimacy, Adultery and Surveillance Technology', *Surveillance and Society*, 11(3): 301–310.

Grimshaw, P. (2002) 'Interracial Marriages and Colonial Regimes in Victoria and Aotearoa/ New Zealand', *Frontiers*, 23(3): 12–28.

Grimshaw, P., et al. (1994) *Creating a Nation*. Melbourne: McPhee Gribble.

Grossi, R. (2012) 'The Meaning of Love in the Debate for Legal Recognition of Same-Sex Marriage in Australia', *International Journal of Law in Context*, 8(4): 487–505.

Grossi, R. (2013) 'Love: A Feminist Conundrum', *The Feminist Wire*, 2 September. www.thefeministwire.com

Grossi, R. (2018) What Has Happened to the Feminist Critique of Romantic Love in the Same-Sex Marriage Debate? In Garcia-Andrade, A., Gunnarsson, L., and Jonasdottir, A. (eds.) *In Feminism and the Power of Love: Interdisciplinary Intervention*. London and New York: Routledge.

The Guardian (2015a) 'Ashley Madison Claims Site Has Plenty of Female Users Eager to Cheat', Sam Thielman, Monday 31 August. www.theguardian.com/technology/2015/aug/31/ashley-madison-claims-site-has-plenty-of-female-users

The Guardian (2015b) 'Ashley Madison Denies Allegations of "Fembot Army"', Alex Hearn, Wednesday 2 September. www.theguardian.com/technology/2015/sept/02/ashley-madison-denies-allegations-of-fembot-army

The Guardian (2015c) 'Ashley Madison's Terms and Conditions Told Users It Ran Fake Accounts', Alex Hearn, Wednesday 9 September. www.theguardian.com/technology/2015/sept/09/ashley-madison-terms-and-conditions-told-users-it-ran-fake-accounts

Gunnarson, L. (2011) 'Love-Exploitable Resource or "No-Lose Situation"? Reconciling Jonasdottir's Feminist Views with Bhaskar's Philosophy of Meta-Reality', *Journal of Critical Realism*, 10(4): 419–441.

Gunnarson, L. (2014a) *The Contradictions of Love: Towards a Feminist-Realist Ontology of socio-sexuality*. London and New York: Routledge.

Gunnarson, L. (2014b) Loving Him for Who He Is: The Microsociology of Power. In Jonasdottir, A.G. and Ferguson, A. (eds.) *Love a Question for Feminism in the Twenty First Century*. London and New York: Routledge.

Gunnarson, L., Andrade, A.G., and Jonasdottir, A.G. (eds.) (2018) The Power of Love: Towards an Interdisciplinary and Multi-Theoretical Feminist Love Studies. In Garcia-Andrade, A., Gunnarsson, L., and Jonasdottir, A. (eds.) *Feminism and the Power of Love: Interdisciplinary Intervention*. London and New York: Routledge.

Hamilton, C. (2012) 'You'll Love the Way It Makes You Feel: The Passion and Privation of Modern Work Culture', *Cultural Studies Review*, 18(3): 411–416.

Hannaford, D. (2015) 'Technologies of the Spouse: Intimate Surveillance in Senegalese Transnational Marriages', *Global Networks*, 15(1): 43–59.

Hansen, M.H. and Pang, C. (2010) Idealizing Individual Choice: Work, Love and Family in the Eyes of Young Rural Chinese. In Hansen, M.H. and Svarverud, R. (eds.) *China: The Rise of the Individual in Modern Chinese Society*. Copenhagen: NIAS.

Harmon, S. (2018) 'Aziz Ansari Responds to Sexual Assault Allegations: Master of None Actor Says He Believed Sexual Activity Was "Completely Consensual"', *The Guardian*,

15 January. www.the guardian.com/culture/2018/jan/15/aziz-ansari-responds to sexual-assault-allegation

Harrell, E. (2008) 'Are Romantic Movies Bad For You?', *Time*, 23 December. www.content.time.com

Harvey, D. (1991) *The Condition of Postmodernity: The Enquiry into the Origins of Cultural Change*. New Jersey: Wiley-Blackwell.

Hirsch, J.S. and Wardlow, H. (2006) *Modern Loves: The Anthropology of Romantic Courtship and Companionate Marriage*. Michigan: University of Michigan Press.

Hochschild, A.R. (1979) 'Emotion Work, Feeling Rules and Social Structure', *American Journal of Sociology*, 85(3): 551–575.

Hochschild, A.R. (2003) *The Managed Heart: Commercialization of Human Feelings*. Berkeley, CA: University of California Press.

Holmes, M. (2015) 'Men's Emotions: Heteromasculinity, Emotional Reflexivity and Intimate Relationships', *Men and Masculinities*, 18: 176–192.

Horst, H. (2006) 'The Blessings and Burdens of Communication: Cell Phones in Jamaican Transnational Social Fields', *Global Networks*, 6(2): 143–159.

Hughes, J. (2010) 'Emotional Intelligence: Elias, Foucault and the Reflective Emotional Self', *Foucault Studies*, 8: 28–52.

Ikels, C. (2004) *Filial Piety, Practice and Discourse in Contemporary Asian Society*. Stanford, CA: Stanford University Press.

Illouz, E. (1997) *Consuming the Romantic Utopia: Love and the Cultural Contradictions of Capitalism*. Berkeley, CA: University of California Press.

Illouz, E. (1998) 'The Lost Innocence of Love: Romance as a Postmodern Condition', *Theory, Culture and Society*, 15(3–4): 161–186.

Illouz, E. (2003) *Oprah Winfrey and the Glamour of Misery: An Essay on Popular Culture*. New York: Columbia University Press.

Illouz, E. (2007) *Cold Intimacies: The Making of Emotional Capitalism*. Cambridge: Polity Press.

Illouz, E. (2010) 'Love and Its Discontents: Irony, Reason, Romance', *The Hedgehog Review*, 12(1), Spring: 1–15.

Illouz, E. (2012) *Why Love Hurts: A Sociological Explanation*. Cambridge: Polity Press.

Illouz, E. (2014) *Hard-Core Romance: Fifty Shades of Grey, Best Sellers and Society*. Chicago: University of Chicago Press.

Illouz, E. (2016) Marriage Is Not What It Seems Adapted from Stephanie Coontz. In *The Way We Never Were*. New York: Basic Books.

Indecent Proposal (1993) 'Dir Adrian Lyne'. www.youtube.com/watch

Irwin, S. (2005) *Reshaping Social Life*. London and New York: Routledge.

Isbister, G. (2008) 'Sex and the City: A Postfeminist Fairy Tale', Proceedings of 'Sustaining Culture' 2008 Annual Conference of the Cultural Studies Association of Australia. http://unsa.edu.au/com/csaa/onlineproceedings.htm

Jackson, S. (2006) 'Gender, Sexuality and Heterosexuality: The Complexity (and Limits) of Heteronormativity', *Feminist Theory*, 7(1): 105–121.

Jackson, S. (2014) Love, Social Change, and Everyday Heterosexuality. In Jonasdottir, A.G. and Ferguson, A. (eds.) *Love: A Question for Feminism in the Twenty First Century*. London and New York: Routledge.

Jamieson, L. (1998) *Intimacy: Personal Relationships in Modern Societies*. Cambridge: Polity Press.

Jamieson, L. (1999) 'Intimacy Transformed? A Critical Look at the "Pure Relationship"', *Sociology*, 33: 477–494.

Jamieson, L. (2011) 'Intimacy as a Concept: Explaining Social Change in the Context of Globalization or Another Form of Ethnocentrism', *Sociological Research Online*, 16(4): 15. www.scoresonline.org.uk

Janelli, R. and Yim, D. (2004) The Transformation of Filial Piety in Contemporary South Korea. In Inkels, C. (ed.) *Filial Piety, Practice and Discourse in Contemporary East Asia.* Stanford, CA: Stanford University Press.

Jankowiak, W.R. (ed.) (2008) *Intimacies: Love and Sex Across Cultures.* New York: Columbia University Press.

Johnson, P. (2005) *Love, Heterosexuality and Society.* London and New York: Routledge.

Jonasdottir, A.G. (1994) *Why Women are Oppressed.* Philadelphia: Temple University Press.

Jonasdottir, A.G. (2011) What Kind of Power Is 'Love Power'? In Jonasdottir, A.G., Bryson, V., and Jones, K.B. (eds.) *Sexuality, Gender and Power: Intersectional and Transnational Perspectives.* New York: Routledge.

Jonasdottir, A.G. (2014) Love Studies: A (Re) New(ed) Field of Knowledge Interests. In Jonasdottir, A.G. and Ferguson, A. (eds.) *Love: A Question of Feminism in the Twenty-First Century.* New York and London: Routledge.

Jonasdottir, A.G. and Ferguson, A. (eds.) (2014) *Love: A Question of Feminism in the Twenty-First Century.* New York and London: Routledge.

Julien, D., Chartrand, E., Simard, M.C., Bouthillier, D., and Begin, J. (2003) 'Conflict, Social Support and Relationship Quality: An Observational Study of Heterosexual, Gay Male and Lesbian Couples' Communication', *Journal of Family Psychology,* 17: 419–428.

Kiernan, A. (2006) No Satisfaction: Sex and the City, Run Catch Kiss, and the Conflict of Desires in Chick Lit's New Heroines. In Ferriss, S. and Young, M. (eds.) *Chick Lit: The New Woman's Fiction.* New York: Routledge.

Kinsey, A., Pomeroy, W., and Martin, C. (1948). *The Kinsey Report: Sexual Behaviour in the Human Male.* Philadelphia: Saunders.

Kinsey, A., Pomeroy, W., Martin, C., and Gebhard, P.H. (1953). *The Kinsey Report: Sexual Behaviour in the Human Female.* Philadelphia: Saunders.

Kipnis, L. (2003) *Against Love: A Polemic.* New York: Pantheon.

Krutnick, F. and Neale, S. (1990) *Popular Film and Television Comedy.* London: Routledge.

Kurdeck, L.A. (2006) 'Differences between Partners from Heterosexual, Gay and Lesbian Cohabiting Couples', *Journal of Marriage and the Family,* 68: 509–528.

Lacan, J. (1977) *Ecrits, A Selection* (Jacques Lacan). Trans., Alan Sheridan. New York: Norton.

Lake, M. (2013) 'Women's and Gender History in Australia: A Transformative Practice', *Journal of Women's History,* 25(5): 190–211.

Lan, P.C. (2006) *Global Cinderella: Migrant Domestic and Newly Rich Employers in Taiwan.* Durham, NY: Duke University Press.

Langford, W. (1999) *Revolutions of the Heart: Gender, Power and the Delusions of Love.* London and New York: Routledge.

Langhamer, C. (2007) 'Love and Courtship in Mid-Twentieth Century England', *The Historical Journal,* 50: 196.

Lawrence, D.H. (1928) (2001) *Lady Chatterley's Lover.* New York: Modern Library Classics.

Lemmings, D. and Brooks, A. (eds.) (2014) *Emotions and Social Change: Historical and Sociological Perspectives.* London and New York: Routledge.

Leonard, S. (2009) *Fatal Attraction.* New York: Wiley-Blackwell.

Leonard, S. (2010a) I Hate My Job, I Hate Everybody Here. In Negra, D. and Tasker, Y. (eds.) *Interrogating Postfeminism: Gender and the Politics of Popular Culture.* Durham, NC: Duke University Press.

Leonard, S. (2010b) '"That's All I Intend to Share Right Now": Adultery and Privacy in the Good Wife', Flow TV, 2 March. www://flowtv.org/2010/07/that's-all-I-intend-to-share-right-now/

Li, Y. (2008) Chinese Women's Studies of Love, Marriage and Sexuality. In Jackson, S., Liu, J., and Woo, J. (eds.) *East Asian Sexualities: Modernity, Gender and New Cultures.* London and New York: Zed Books.

Liebelt, C. (2011) *Caring for the Holyland: Filipina Domestic Workers in Israel*. New York and Oxford: Berghahn Books.

Livingston, G. (2015) 'For Most Highly Educated Women, Motherhood Doesn't Start Until the 30s', *Pew Research Center*. www.pewresearch.org/2015/for-most-highly-educated-women-motherhood-doesn'tstart-until-the30s

Lohmeirer, C. (2012) 'Book Review Melissa Gregg's *Work's Intimacy*', *New Media and Society*, 14(7): 1244–1245.

Lomas, T, Cartwright, T, Edgington, T., and Ridge, D. (2016) 'New Ways of Being a Man': "Positive" Hegemonic Masculinity in Meditation Based Communities of Practice', *Men and Masculinities*, 19(3): 289–310.

Lugu Lake Mosua Cultural Development Association (2006) *Walking Marriages*. www.mosuoproject.org/walking.htm

Luhmann, N. (1986) *Love as Passion: The Codification of Intimacy*. Cambridge, MA: Harvard University Press.

Luther, J. (2013) 'Beyond Bodice-Rippers: How Romance Novels Came to Embrace Feminism', *The Atlantic*, 18 March. www.theatlantic.com

Mabry, A.R. (2006) About a Girl: Female Subjectivity and Sexuality in Contemporary "Chick" Culture. In Ferriss, S. and Young, M. (eds.) *Chick Lit: The New Woman's Fiction*. New York: Routledge.

Madianou, M. and Miller, D. (2012) *Migration and New Media: Transnational Families and Polymedia*. London and New York: Routledge.

Maia, S.M. (2007) Brazilian Exotic Dancers in New York: Desire and National Identity. PhD Thesis: The Graduate Centre, City University of New York (CUNY).

Manzerolle, V. and Kjosen, A. (2012) 'The Communication of Capital: Digital Media and the Logic of Acceleration', *Triple C*, 10(2): 214–229.

McAleer, J. (1999) *Passion's Fortune: The Story of Mills and Boon*. Oxford: Oxford University Press.

McAlister, J. and Teo, H.-M. (2017) Love in Australian Romance Novels. In Teo, H.-M. (ed.) *The Popular Culture of Romantic Love in Australia*. Melbourne: Australian Scholarly Publishing, 194–222.

McCabe, J. (2009) Lost in Transition: Problems of Modern (Heterosexual) Romance and the Catatonic Male Hero in Post-Feminist Age. In Abbott, S. and Jermyn, D. (eds.) *Falling in Love Again: Romantic Comedy in Contemporary Cinema*. New York: IB Tauris and Co, Ltd., 160–175.

McDonald, J. (2009) 'Homme-Com': Engendering Change in Contemporary Romantic Comedy. In Abbott, S. and Jermyn, D. (eds.) *Falling in Love Again: Romantic Comedy in Contemporary Cinema*. New York: IB Taurus and Co, Ltd., 146–159.

McGrath, A. (2002) 'White Brides: Images of Marriage across Colonizing Boundaries', *Frontiers*, 23(3): 76–108.

Mody, P. (2008) *The Intimate State: Love, Marriage and the Law in Delhi*. New Delhi: Routledge.

Montes, V. (2013) 'The Role of Emotions in the Construction of Masculinity: Guatemalan Migrant Men, Transnational Migration, and Family Relations', *Gender and Society*, 27(4): 469–490.

Mortimer, C. (2010) *Romantic Comedy*. London and New York: Routledge.

Morton, A. (2018a) 'Meaghan and Mrs Simpson', *The Daily Telegraph Magazine*, 17 February: 14–19.

Morton, A. (2018b) *Wallis in Love: The Untold True Passion of the Duchess of Windsor*. London: Michael O'Mara Books.

Mullen, P.E. (1991a) 'The Pathology of Passion', *The British Journal of Psychiatry*, 158: 593–601.

Mullen, P.E. (1991b) 'Paul Mullen: A Celebration', *Criminal Behaviour and Mental Health*, 20(3): 161–164.

Mullen, P.E. (2010) 'Paul Mullen: A Celebration,' *Criminal Behaviour and Mental Health*, 20(8): 161–164.

Mullen, P.E. (2018) 'Jealousy: The Pathology of Passion', *The British Journal of Psychiatry*, 158(5): 593–601.

Mulvey, L. (1975) 'Visual Pleasure and Narrative Cinema', *Screen*, 16(3): 6–18.

Mulvey, L. and Backman Rogers, A. (2015) *Feminism: Diversity, Difference and Multiplicity in Contemporary Film Cultures*. Amsterdam: Amsterdam University Press.

Mulvey, L., Backman Rogers, A., and Van den Oever, A. (2015) 'Feminist Film Studies 40 Years after "Visual Pleasure and Narrative Cinema," a Triologue', *NECSUS*, 4(1): 67–79.

National Geographic (2016) *Walking Marriages*. www://channel.nationalgeographic.com/taboo/videos/walking-marriage

National Union of Students (2010) 'Hidden Marks: A Study of Women Students' Experience of Harassment, Stalking, Violence and Sexual Assault', *National Union of Students News*, 6 May 2010. www.nus.org.uk

National Union of Students (2013) 'That's What She Said: Women Students' Experiences of "Lad Culture" in Higher Education'. www.nus.org.uk

Neale, S. (2009) *Genre and Contemporary Hollywood*. New York and London: Bloomsbury.

Nehring, D. (2013) '*Why Love Hurts* Review', *Sociology*, 47(6): 1233–1239.

Nicholls M. (2014) Marriage, Romance and Mourning Movement in Cherie Nowlan's Thank God He Met Lizzie, *Journal of Popular Romance Studies*, October 2014, http://www.jprstudies.org.

Ong, A. (1999) *Flexible Citizenship: The Cultural Logics of Transnationality*. Durham, NC: Duke University Press.

Ong, A. (2006) *Neoliberalism an Exception: Mutations in Citizenship and Sovereignty*. Durham, NC: Duke University Press.

Oswald, R.F., Blume, L.B., and Marks, S.R. (2005) Decentering Heteronormativity: A Model for Family Studies. In Bengtson, V.I., Acock, A.C., Allen, K.R., Dilworth-Anderson, P., and Klein, D.M. (eds.) *Sourcebook of Family Theory and Research*. Thousand Oaks, CA: Sage.

Oxfeld, E. (2005) Cross-Border Hypergamy? Marriage Exchanges in a Transnational Hakka Community. In Constable, N. (ed.) *Cross-Border Marriages: Gender and Mobility in Transnational Asia*. Philadelphia: University of Philadelphia Press.

Padilla, M. (2007) *Caribbean Pleasure Industry: Tourism, Sexuality and AIDS in the Dominican Republic*. Chicago: University of Chicago Press.

Parrenas, R.S. (2001) *Servants of Globalization: Women, Migration and Domestic Work*. Stanford, CA: Stanford University Press.

Parrenas, R.S. (2005a) *Children of Global Migration: Transnational Families and Gendered Woes*. Stanford, CA: Stanford University Press.

Parrenas, R.S. (2005b) 'Long Distance Intimacy: Class, Gender and International Relations Between Mothers and Children in Filipino Transnational Families', *Global Networks*, 5(4): 317–336.

Parrenas, R.S. (2008) *The Forces of Domesticity: Filipina Migrants and Domesticity*. New York: New York University Press.

Parrenas, R.S. (2014) 'Intimate Labour of Transnational Communication', *Families, Relationships and Societies*, 3(3): 425–442.

Patrick, K. (2017) Intimate Confessions: The Rise and Fall of Romance Comic Books in Australia. In Teo, H.-M. (ed.) *The Popular Culture of Romantic Love in Australia*. Melbourne: Australian Scholarly Publishing, 223–256.

Pearce, L. and Stacey, J. (1995) *Romance Revisited*. New York: New York University Press.

Peiss, K. (2011) *Hope in a Jar: The Making of America's Beauty Culture*. Philadelphia: University of Pennsylvania Press.

Peplau, L.A. (2001) 'Rethinking Women's Sexual Orientation: An Interdisciplinary Relationship Focused Approach', *Personal Relationships*, 8: 1–19.

Peplau, L.A. and Fingerhut, A.W. (2007) 'The Close Relationship of Lesbian and Gay Men', *Annual Review of Psychology*, 58: 405–424.

Pew Research Center (2013) Internet and Technology, 'Online Dating and Relationships', M. Duggan and A. Smith, 21 October, 1–18. www.pewresearch.org

Pew Research Center (2014) 'The Next America', 10 April. www.pewresearch.org

Pew Research Center (2015) 'Teens, Technology and Romantic Relationships: From Flirting to Breaking Up, Social Media and Mobile Phones Are Woven into Teens Romantic Lives'. www.pewinternet.org/2015/10/01/teens-technology-and-romantic-relationships

Phipps, A. and Young, I. (2014) 'Neoliberalism and "Lad Culture" in Higher Education', *Sexualities*, 49(2): 305–322. www.journals.sagepub.com

Phipps, A. and Young, I. (2015) '"Lad Culture" in Higher Education: Agency in the Sexualization Debates', *Sexualities*, 18(4): 459–479. www.journals.sagepub.com

Piper, N. and Roces, S. (eds.) (2003) *Wife or Worker? Asian Women and Migration*. New York: Rowman and Littlefield.

Platt, M., Yeoh, B.A., Acadera, K.A., Yen, K.C., Baey, G., and Lam, T. (2004) 'Migration and Information Communications Technology Use: A Case Study of Indonesian Domestic Workers in Singapore'. www.solutionexchange-un-gen-gym.net/up-content/uploads/2015/11/Migration-and-ICT-Use.pdf

Pleios, G. (2012) 'Communication and Symbolic Capitalism: Rethinking Marxist Communication Theory in the Light of the Information Society', *Triple C*, 10(2): 230–252.

Preston, J. and Lowenthal, M. (eds.) (1996) *Friends and Lovers: Gay Men Write about the Families they Create*. New York: Plume.

Prins, K.S., Buunk, B.P., and Van Yperen, N.W. (1993) 'Equity, Normative Disapproval and Extramarital Relationships', *Journal of Social and Personal Relationships*, 10: 39–53.

Purcell, C. (2013) 'Fifty Shades of Feminism: A Response to E. L. James Fifty Shades of Grey', *Huffington Post*, 2 January.

Quah, S.R. (2008) *Families in Asia: Home and Kin*. London and New York: Routledge.

Rayner, J. (2017) Romantic Love in the Australian Cinema. In Teo, H.-M. (ed.) *The Popular Culture of Romantic Love in Australia*. Melbourne: Australian Scholarly Publishing, 171–193.

Richtel, M. (2012) 'Young, in Love and Sharing Everything, Including a Password', *The New York Times*, 18 January. http://nytimes.com/2012/01/18/us/teenagers-sharing-passwords-as-show-of-affection.html

Roberts, S. (2013) 'Boys Will Be Boys . . . Won't They? Change and Continuities in Contemporary Working-Class Masculinities', *Sociology*, 47: 671–686.

Roces, M. (2003) Sisterhood Is Local: Filipino Women in Mount Isa. In Piper, N. and Roces, S. (eds.) *Wife or Worker? Asian Women and Migration*. New York: Rowman and Littlefield.

Roseneil, S. (2000) 'Queer Framework and Queer Tendencies: Towards and Understanding of Postmodern Transformations of Sexuality', *Sociological Research Online*, 5(3). www.socresearchonline.org.uk

Roseneil, S. and Budgeon, S. (2004) 'Cultures of Intimacy and Care Beyond the Family: Personal Life in the Early 21st Century', *Current Sociology*, 52: 135–159.

Rosenfeld, M.J. and Thomas, R.J. (2012) 'Searching for a Mate: The Rise of the Internet as a Social Intermediary', *American Sociological Review*, 77(4): 523–547.

Rowntree, M.R. (2015) 'Feminine Sexualities in the Chick Genre', *Feminist Media Studies*, 15(3): 508–521.

Rusu, M.S. (2018) 'Theorising Love in Sociological Thought: Classical Contributions to a Sociology of Love', *Journal of Classical Sociology*, 18(1): 13–20.

Salvin, S. (2009) '"Instinctively, I'm Not Just a Sexual Beast": The Complexity of Intimacy Among Australian Gay Men', *Sexualities*, 12(1): 79–96.

Shamalzbauer, L. (2005) 'Searching for Wages and Mothering from Afar: The Case of Honduran Transnational Families', *Journal of Marriage and Family*, 66: 1217–1231.

Shen, H.H. (2005) '"The First Taiwanese Wives" and the "Chinese Mistresses": The International Division of Labour in Familial and Intimate Relations across the Taiwan Strait', *Global Networks*, 5(4): 419–437.

Shen, S.S. (2008) 'The Purchase of Transnational Intimacy: Women's Bodies, Transnational Masculine Privileges in Chinese Economic Zones', *Asian Studies Review*, 32: 57–75.

Shih, S.M. (1999) Gender and Geopolitics of Desire: The Seduction of Mainland Women in Taiwan and Hong Kong Media. In Yang, M.M. (ed.) *Spaces of Their Own*. Minneapolis: University of Minneapolis Press, 278–307.

Shumway, D. (2003) *Modern Love: Romance, Intimacy and the Marriage Crisis*. New York: Routledge.

Sim, A.S.C. and Wee, V. (2009) 'Undocumented Indonesian Workers: The Human Outcome of Colluding Interests', *Critical Asian Studies*, 41: 165–188.

Simon, H.A. (1990) 'Invariants of Human Behaviour', *Annual Review of Psychology*, 41: 1–20.

Slater, D. (2013) *Love in the Time of Algorithms: What Technology Does to Meeting and Mating*. New York: Current Books.

Slavin, S. (2009) '"Instinctively, I'm Not Just a Sexual Beast": The Complexity of Intimacy Among Australian Gay Men', *Sexualities*, 12(1): 79–96.

Smart, C. (2007) *Personal Life: New Directions in Sociological Thinking*. Cambridge: Polity Press.

Smart, C. and Shipman, B. (2004) 'Visions in Monochrome: Families, Marriage and the Individualization Thesis', *British Journal of Sociology*, 55: 491–509.

Sorokin, P.A. (1948) *The Reconstruction of Humanity*. Boston, MA: Beacon Press.

Sorokin, P.A. (1950) *Altruistic Love: A Study of American 'Good Neighbours' and Christian Saints*. Boston, MA: Beacon Press.

Sorokin, P.A. (1954) *The Ways and Power of Love: Types, Factors and Techniques of Moral Transformation*. Boston, MA: Beacon Press.

Stearns, P.N. (1989) *Jealousy, the Evolution of an Emotion in American History*. New York: New York University Press.

Stearns, P.N. (2006) *The Causes and Consequences of High Anxiety*. New York: Routledge.

Summers, A. (1975) *Damned Whores and God's Police: The Colonization of Women in Australia*. Ringwood, VIC: Penguin Books.

Swidler, A. (2001) *Talk of Love: How Culture Matters*. Chicago: University of Chicago Press.

Tasker, Y. and Negra, D. (2005) 'In Focus: Postfeminism and Contemporary Media Studies', *Cinema Journal*, 44(2): 107–110.

Teo, H.-M. (2006) The Americanisation of Romantic Love in Australia. In Curthoys, A. and Lake, M. (eds.) *Connected Worlds: History in Transnational Perspective*. Canberra: ANU E-Press, 171–192.

Teo, H.-M. (2014a) '"Introduction" Special Issue: The Popular Culture of Romantic Love in Australia', *Journal of Popular Romance Studies*, 4(2).

Teo, H.-M. (2014b) '"We Have to Learn to Love Imperially": Love in Late Colonial and Federation Australian Romance Novels', *Journal of Popular Romance Studies*, 4(2).

http://jprstudies.org/2014/10/we-have-to-learn-to-love-imperialy-love-in-late-colonial-and-federation-australian-romance-novelsby-hsu-ming-teo

Teo, H.-M. (ed.) (2017a) *The Popular Culture of Romantic Love in Australia*. Melbourne, VIC: Australian Scholarly Publishing.

Teo, H.-M. (2017b) Introduction: The Popular Culture of Romantic Love in Australia. In Teo, H.-M. (ed.) *The Popular Culture of Romantic Love in Australia*. Melbourne, VIC: Australian Scholarly Publishing, 1–39.

Thagaard, T. (1997) 'Gender, Power and Love: A Study of Interaction Between Spouses', *Acta Sociologica*, 40(4): 357–376.

Thai, H.C. (2005) Clashing Dreams in the Vietnamese Diaspora: Highly Educated Overseas Brides and Low-Wage Husbands. In Constable, N. (ed.) *Cross-Border Marriages: Gender and Mobility in Transnational Asia*. Philadelphia: University of Philadelphia Press.

Thai, H.C. (2008) *For Better or Worse: Vietnamese International Marriages in the New Global Economy*. New Brunswick, NJ: Rutgers University Press.

Thompson, D. (2014) 'Oscar Voters: 94% White, 76% Men, and an Average of 63 Years Old', *The Atlantic*, 2 March. www.theatlantic.com/entertainment/archive/2014/03/oscar-voters-94-white-76-men-and-an-average-of-63-years-old

Thormeer, M.C., Umberson, D., and Pudrovska, T. (2013) 'Marital Processes around Depression: A Gendered and Relational Perspective', *Society and Mental Health*, 3: 151–169.

Toye, M. (2010) 'Towards a Poethics of Love', *Feminist Theory*, 11(1): 39–55.

Tsapelas, I., Fisher, H.E., and Aron, A. (2010) Infidelity: When, Where, Why. In Cupach, W.R. and Spitzberg, B.H. (eds.) *The Dark Side of Close Relationships*. New York: Routledge, 175–196.

Turner, B. (1998) 'Consuming the Romantic Utopia: Love and the Cultural Contradictions of Capitalism by Eva Illouz', *Body and Society*, 4: 115–117.

Turner, D. (2002) *Fashioning Adultery: Gender, Sex and Civility in England, 1660–1740*. Cambridge: Cambridge University Press.

Umberson, D., Thomeer, M.B., and Lodge, A.C. (2015) 'Intimacy and Emotion Work in Lesbian, Gay and Heterosexual Relationships', *Journal of Marriage and Family*, 77: 542–556.

Unfaithful (2002, dir Adrian Lyne). www.youtube.com/watch

Van den Eijnden, R.J.J., Bunk, B.P., and Bosveld, W. (2000) 'Feeling Similar or Feeling Unique: How Men and Women Perceive Their Own Sexual Behaviours', *Personality and Social Psychology Bulletin*, 26: 1540–1549.

Wang, H.Z. and Chang, S.M. (2002) 'The Commodification of International Marriage: Cross-Border Marriage Business between Taiwan and Vietnam', *International Migration*, 40: 93–116.

Way, K. (2018) 'I Went on a Date with Aziz Ansari: It Turned into the Worst Night of My Life', *babe.net*. https://babe.net/2018/01/13/aziz-ansari-28355

Weber, M. (1946) Religious Rejection of the World and their Directions. In Gerth, H.H. and Wright Mills, C. (eds.) *From Max Weber: Essays in Sociology*. New York: Oxford University Press, 323–359.

Weeks, J. (2000) *Making Sexual History*. Oxford: Polity Press.

Weeks, J., Heaphy, B., Donovan, C., and Carr, L. (2001) *Same Sex Intimacies Families of Choice and Other Life Experiments*. London and New York: Routledge.

Weeks, J. (2007) *The World We Have Won: The Remaking of Erotic and Intimate Life*. London: Routledge.

West, C. and Zimmerman, D.H. (2009) 'Accounting for Doing Gender', *Gender and Society*, 23: 112–122.

Whelehan, I. (2000) The Bridget Jones Effect. In Whelehan, I. (ed.) *Overloaded: Popular Culture and the Future of Feminism*. London: Women's Press, 135–153.

Willmot, H. (2007) 'Young Women, Routes through Education and Employment and Discursive Constructions of Love and Intimacy', *Current Sociology*, 55(3): 446–466.

Wollstonecraft, M. (2004) *A Vindication of the Rights of Woman*. London: Penguin.

Yan, Y. (2000) Introduction: Conflicting Images of the Individual and Contested Process of Individualization. In Hansen, M.H. and Svarverud, R. (eds.) *China and the Rise of the Individual in Modern Chinese Society*. Copenhagen: NIAS.

Yates, C. (2007) *Masculine Jealousy and Contemporary Cinema*. London: Palgrave Macmillan.

Yates, C. (2009) Masculinity, Flirtation and Political Communication in the UK. In Sclater, S.D., Jones, D.W., Price, H., and Yates, C. (eds.) *Emotion*. London: Palgrave Macmillan.

Yates, C. (2016) 'Masculinity, Jealousy and Cinema', Unpublished Lecture. Bournemouth University.

Yeoh, B.S.A., Huang, S., and Lam, T. (2005) 'Transnationalizing the "Asian" Family: Imaginaries, Intimacies, and Strategic Intents', *Global Networks*, 5: 307–315.

Yossman, K. (2018) 'Rise of the "Sapiosexual": Why A-List Men Date Brainy Women', *The Daily Telegraph*, 13 April, 2. www.telegraph

Zelizer, V. (2005) The Purchase of intimacy. Princeton: Princeton university press.

INDEX